1/28/94

D0983775

Evolution and the Recognition Concept of Species

HUGH E. H. PATERSON

Evolution and the Recognition Concept of Species *Collected Writings*

Edited by Shane F. McEvey

The Johns Hopkins University Press
Baltimore and London

© 1993 The Johns Hopkins University Press
All rights reserved
Printed in the United States of America on acid-free paper

The Johns Hopkins University Press
2715 North Charles Street
Baltimore, Maryland 21218-4139
The Johns Hopkins Press Ltd., London

LIBRARY OF CONGRESS CATALOGING-IN-PUBLICATION DATA

Paterson, H. E. H.
  Evolution and the recognition concept of species : collected writings /
Hugh E. H. Paterson : edited by Shane F. McEvey.
      p.   cm.
  Includes bibliographical references and indexes.
  ISBN 0-8018-4409-6
  1. Species.   2. Evolution (Biology).   I. McEvey, Shane F.   II. Title.
QH380.P38   1992
575—dc20   92-4779

To my graduate students
past and present

Contents

Editor's Preface   ix

Author's Preface   xiii

1. The Term "Isolating Mechanisms" as a Canalizer of Evolutionary Thought   1
2. The Role of Postmating Isolation in Evolution   11
3. More Evidence against Speciation by Reinforcement   20
4. A Comment on "Mate Recognition Systems"   32
5. The Continuing Search for the Unknown and Unknowable: A Critique of Contemporary Ideas on Speciation   35
6. Epitaph to a Scientific Gadfly   58
7. Morphological Resemblance and Its Relationship to Genetic Distance Measures   63
8. Perspective on Speciation by Reinforcement   76
9. Darwin and the Origin of Species   92
10. The Recognition Concept of Species: Macnamara Interviews Paterson   104
11. Simulation of the Conditions Necessary for the Evolution of Species by Reinforcement   124
12. The Recognition Concept of Species   136
13. Environment and Species   158
14. On Defining Species in Terms of Sterility: Problems and Alternatives   168
15. A View of Species   180
16. Updating the Evolutionary Synthesis   194
17. The Recognition of Cryptic Species among Economically Important Insects   199

Complete Bibliography of H.E.H. Paterson   213

Supplemental References   219

Taxonomic Index   221

Author Index   225

Subject Index   229

Editor's Preface

Paterson's ideas on species and speciation have become prominent in the study of evolution. Unlike earlier evolutionists, who wrote influential books, Paterson has introduced his concept through a series of articles, of which some have had only limited circulation. Consequently it has been difficult for many to appreciate the whole and to understand that no single article represents an adequate statement of Paterson's overall view. To rectify this situation is one reason for this book; the other is to celebrate Paterson's sweeping contribution to evolutionary biology on the occasion of his retirement. This volume is a collection of all of Paterson's work on species and speciation.

Perusal of his complete bibliography, additionally included, shows how Paterson's theories rest substantially on extensive experience in ecological genetics, taxonomy, entomology, ornithology, and parasitology. Also evident in the bibliography is the extension of his idea in works largely by his students. See for example: Robertson and Paterson (1982), Lambert and Paterson (1984), and Masters, Lambert, and Paterson (1984). When all Paterson's papers are viewed in chronological sequence, one sees clearly the development of his thinking from the time of facing a number of basic biological problems (1953–1975) to a conceptual revelation (1973–85, chaps. 1–12) and, ultimately, to its far-reaching implications (1986–91, chaps. 13–17). This volume picks up the story in 1973 and 1975 and concludes in 1991 with his most recent work. To link the articles and to explain what circumstances lay behind them, Paterson has written introductions to each chapter.

Every attempt has been made to reproduce his papers in this book without change, although typographical errors detected in the originals have been corrected and American spelling and punctuation have been adopted throughout. More difficult to deal with are situations where a mistake in the original text led to an unintended meaning. Fortunately there are few such errors, and only two are significant: "and other" is changed to "another" in Lambert and Paterson (1982:293) (chap. 7); and "preference" is changed to "reference" in Paterson (1982:54) (chap. 8). Some errors have been detected in citations and bibliographic detail; in collaboration with the author these have been corrected. The callout number for reference 26 is inserted at the appropriate place in Paterson (1978:370) (chap. 3); the reference callout number 12 is changed to 18 and reference 32 is linked to Fisher (1958) in Paterson (1981, pp. 115 and 116, respectively) (chap. 5). Where a publication was originally cited as "in press," the full reference has been added.

Citations used in the Author's Preface and introductory paragraphs are collected in a supplemental reference section at the end of the book. Indexes to subjects, species and genera, and authors cited are also presented. To facilitate accurate citation, the original pagination is embedded in the text within square brackets and in an alternate font. As a convenience to readers, original references to those papers presented in this volume are supplemented by reference to the appropriate chapter in brackets.

Eight papers from the *South African Journal of Science* have been reprinted in this book, and for this I am most grateful for the full cooperation and support received from its editor, Dr. Graham Baker. Acknowledgments are made also to the Transvaal Museum, the *Journal of the Royal Society of Western Australia, Evolution, Evolutionary Theory, Pacific Science, Rivista di Biologia–Biology Forum, Cold Spring Harbor Symposia of Quantitative Biology,* Springer-Verlag and Edinburgh University Press for granting permission to reprint material.

Acknowledgment to the institutions at which Paterson (and his coauthors) worked when these papers were written, excluded from the chapters, is summarized as follows: Department of Zoology, University of Western Australia (1967–1975); Department of Zoology, University of the Witwatersrand, Johannesburg (1975–1985 [with Marc Centner 1984]); and Department of Entomology, University of Queensland, Brisbane (1985–present). While at Wits Paterson was professor and head of the Department of Zoology; he was supported by the university's Council Research Committee, and together with his students he worked in the George Evelyn Hutchinson Laboratory. Michael Macnamara was professor and head of the Department of Philosophy at the University of South Africa when he interviewed Paterson in 1984 (chap. 10). David Lambert moved from the Hutchinson Lab to the Department of Zoology, University of Auckland, New Zealand, in 1981.

Proofreading has been especially important; I am very grateful for help from Maureen Coetzee, Richard Hunt, and Shirley Paterson. Max Moulds ably assisted me in the facsimile reproduction of figures.

My colleagues and students in Paris and Sydney constantly requested Paterson's papers and questioned me on the recognition concept; this, together with the cherished memories I have of four years in the Hutchinson Lab (1982–85) and four years in Paris (1985–1990), provided me with the incentive to have this collection published. I owe a special debt of gratitude to Jean David and many other colleagues at the C.N.R.S. Laboratoire de Biologie et Génétique Evolutives (Gif-sur-Yvette), with whom I enjoyed many stimulating discussions on species and speciation.

In the preparation of this book I have been greatly assisted by a number of students in the Department of Plant Pathology and Agricultural Entomology and at Wesley College, University of Sydney. In particular I thank Graeme Barden for checking texts and Matthew and Peter Spencer for help in hardcopy scanning to produce electronic text. I am indebted to Fred McDonald for giving me the opportunity to lecture to his students on species biology, thereby finding, once again and in a different country, a great need for this book.

This work was completed at the Department of Entomology, Australian Museum, Sydney, where I hold a Visiting Research Fellowship.

Shane F. McEvey

Author's Preface

The theoretical views in this book stemmed from eleven years' experience as a medical entomologist, during which time I worked on the systematics and ecology of Diptera of medical importance and the epidemiology of a number of diseases such as arboviruses, malaria, and plague. In particular I specialized in elucidating complexes of sibling species in the genera *Musca, Chrysomya, Culex,* and *Anopheles.* But earlier, from my country boyhood in Natal, I had a deep interest in and knowledge of birds and a broad experience with animals in general. During my professional life I had been an ardent advocate of the biological species concept, as can readily be seen from my papers of the period.

In a paper on the *Anopheles gambiae* complex in Mauritius (Paterson, 1964) I discussed the possibility of two species being able to coexist, as distinct species are supposed to be able to do (Mayr, 1969), when they differ only in being intersterile. It was easy to show on population genetical grounds that inevitably the least abundant form would rapidly be eliminated. This was contrary to the then current views on species as expounded, for example, by Mayr (1963:551). Later, I argued that the biological species concept should not include populations which are distinguished by intersterility alone (Paterson, 1968:263). This was the first crack that I detected in the biological species concept.

The doubts generated by this insight led me to review the nature of species afresh while at the University of Western Australia. To do this I started with a reconsideration of the nature of sex, noting that sex involved the alternation of meiosis, fertilization, and syngamy (fusion of the half-nuclei to form a diploid zygote following fertilization). I noted that fertilization would not occur on its own; it always involved a system of varying complexity which I called the fertilization system. Over a few days of intense intellectual activity involving the crucial, active participation of C. A. Green, the outline of the recognition concept was hammered out.

Chris Green, Rod Mahon, Simon Miles, Jenny Blackwell, Rosemary Irving-Bell, Nick Monzu, and Jeroen den Hollander were all working on evolutionary problems under my supervision. Our exhilaration from these studies and our theorizing were enhanced by the general background of excitement from the unprecedented and unrepeatable scientific revolution which we were living through. Week by week molecular biology was yielding new insights; the story of plate tectonics was unfolding; Lynn Margulis was stretching our imaginations with her illumination of the previously blank Precambrian ages and of the evolution of eukaryotic cells; Schopf's microfossils in the most ancient rocks, including those of Western Australia and the Northern Territory, were supporting Margulis's views; electron microscopy showed us what we had never before seen in cells; we were gaining understanding of proteins and their evolution; Eldredge and Gould were revitalizing paleontology with their interpretations of fossil patterns over time; and electrophoresis was changing our views on natural variation, besides proving a powerful practical tool for our species studies.

In preparation for the Australian and New Zealand Association for the Advancement of Science (ANZAAS) meeting in Perth in 1973, I was asked to write an account of the evolutionary genetic studies of animals being carried out in Western Australia. This provided the opportunity to include some hints of our changed view of species. We were so fascinated with the beauty and simplicity of the concept that we felt that more than a hint of it would lead to its being picked up and disseminated before we had developed it. Quaintly, we really felt that colleagues would at once appreciate the concept's power and would immediately cease speaking of reproductive isolation and isolating mechanisms, and that our role in the revolution would never be noticed! In fact, there was no cause to worry. Our fears about the idea conquering all before it with its compelling simplicity and elegance could not have been more wrong. Twenty years later it still has some conquering to do, but in 1973 it made no impression at all. I don't believe I ever had a reprint request for it. Certainly none of my scientific colleagues ever mentioned it to me.

The paper is a curious mixture of the isolation concept and the recognition concept, reflecting, perhaps, the lingering influences of the biological species concept that had served me well for nearly twenty years. However, I was also following a cautious plan not to rush things.

Much of the paper relates to its stated aim: the provision of an account of the state of animal species studies in Western Australia in 1973. For this reason only some of it is relevant to this book's theme, and that is why I have preferred not to reprint it here. However, because of its obscurity I shall summarize the significant points that were made. This will enable the present compilation to be used to trace the trajectory of the idea of the recognition concept from 1973 onward.

My contribution began under the ambiguous heading "The Species Concept," meaning, I think, the genetical concept. "At the outset it is necessary to outline what is meant by the genetical concept of the species, because the species in genetics does not always correspond to any species currently recognized in taxonomy." This was a reference to the problem of sibling species. It was also an early expression of my views on species in taxonomy and evolutionary genetics which is dealt with in later papers.

Early on I distinguished sex in eukaryotes from the "alternatives to sex" (Haldane, 1955) to be found among the prokaryotes. I then went on: "In sexually reproducing organisms all reproductively mature members of a species can contribute to, and share in its gene pool. Furthermore, the gene pool of this species is isolated from the gene pools of other species due to the functioning of genetically determined isolating mechanisms." This would hardly have caused anyone any surprise at the time. But, in the next paragraph I start subverting this established position: "These ideas can be expressed in other ways. For example, the essential characteristic of an animal species is that it comprises a number of individuals each of which is able to recognize reproductively mature mating partners of the same species. This is achieved by means of a [specific], genetically determined, coadapted behavioral signaling system. Members of a particular species, because of this system, do not usually respond effectively to the analogous signals of members of other species under natural conditions."

My interest in species arose from my curiosity about the causes of diversification of life on earth. At several points in the article attention was focused on the relationship between species and speciation and diversity and ecology.

> The explosive diversification of forms of organisms which apparently followed the evolution of eukaryotic cells in late Precambrian times cannot be attributed to the newly evolved methods of generating genetic diversity alone. Recombination will generate variation, but will not on its own lead to the radiation of forms which we observe. Radiation will, however, be initiated if recombination is restricted to occurring within discrete gene pools. This raises the key question how did the species as an evolutionary phenomenon first arise? No general answer to this question will be attempted because it is very likely that species arose on several independent occasions following the independent acquisition of sexual reproduction by several different eukaryotic Protista (Margulis, 1970). In [this] discussion only the animal lineage will be considered. Most animals are dioecious or can be shown to have had dioecious ancestors. This suggests that the whole animal kingdom evolved from an early dioecious protistan ancestor.

At this point my attempt to insinuate the positive, "recognition," as opposed to the negative, "isolation," approach becomes evident. Of course, today I find this caution irritating. I did not emphasize points as I would have in a direct approach.

Because we are discussing the origin of the species as an evolutionary phenomenon, such an ancestral form must have lacked an isolating mechanism as such. However, being dioecious it must have had a genetically-determined mechanism which enabled the one mating type to recognize the other. Credibly, this system could have provided the basis for the first isolating mechanism which brought the first two animal species into existence, because the principles of allopatric speciation long advocated by Mayr (1942, 1963) are as applicable to this basic system as to more advanced ones. In simple terms the mechanism behind allopatric speciation may be as follows. Once an extrinsic barrier has split the original "gene pool" into two, the genetic structure of each will begin to deviate in response to selection from its distinct environment. Some of the gene substitutions brought about in this way will have a pleiotropic effect on the mate selection mechanism. In this way the mechanisms of the two populations will progressively diverge to the point when the signals of members of the one population are no longer effectively recognized by members of the other. At this stage speciation will have been achieved. An example of a gene with pleiotropic effects of the sort invoked is *yellow* in *Drosophila melanogaster* Meigen. When compared with "wild type" flies homozygotes for this allele are more resistant to starvation (Kalmus, 1941) and the males have a modified courtship pattern (Bastock, 1956). Following effective long term isolation, the mate recognition systems of the two sub-populations may, thus, also function as premating isolating mechanisms, which effectively protect the integrity of their gene pools. This in turn enables the two populations to continue to diverge adaptively even though their ranges may come to overlap.

Details of the speciation scheme outlined above will be disputed by some, but if it happens to be correct it will be noted to have the following implications. Species come into existence as by-products of adaptive evolution. They are not products of selection for diversity, though selection may decide whether a newly evolved species will survive for long. Diversity is not selected for as such, and it is ultimately dependent on factors which favor speciation. However, selection does decide the specific pattern of diversity which comes into existence in a particular environment at a particular time.

Today, of course, I am not entirely happy with this account of speciation. For example, it was incorrect to state that the mate recognition system could have provided the basis for the first isolating mechanism. At best this would have been an isolating *effect*. Furthermore, stating that "the mate recognition systems of the two sub-populations may . . . also function as premating isolating mechanisms, which effectively protect the integrity of their gene pools" demonstrates that I had incompletely absorbed the lessons that George Williams taught in 1966: I did say "*effectively* protect," but I should have avoided using "function."

The last paragraph quoted illustrates that fundamental consequences were stated, but in such a nonconfrontational way as to be almost cryptic. To emphasize what I was saying I could have contrasted my statement with statements to the contrary by Mayr (1949:284): "Speciation is thus an adaptive process toward the most efficient utilization of the environment," or Dobzhansky (1950:415): "The process of speciation must, then, be regarded as an evolutionary adaptation which permits the development of immense organic diversity, particularly the diversity of sympatric species." Dobzhansky, much later (1976:104), reiterated this viewpoint. I emphasized clearly that I believed species to be an incidental effect of adaptation to a new environment.

At this stage I had not fully appreciated the strong relationship between the mate recognition system, and, more generally, the fertilization system, to the organisms' environment and "way of life." I still thought of speciation in terms of pleiotropy in the way that Mayr and Futuyma do today. The paper which helped me to see the inadequacies of assigning a major role to pleiotropy, and which demonstrated the dominating role of the environment in shaping fertilization systems, was Morton (1975).

The section headed "The Study of Isolating Mechanisms" opened with the following paragraph before mentioning a number of students whose studies were relevant. It only hints that, if allopatric speciation is the usual mode of speciation, there is a problem with the term "isolating mechanisms."

> It is generally accepted that postmating isolating mechanisms arise as pleiotropic by-products of adaptive evolution in isolated populations. This is because it is not easy to see how they could be evolved directly by natural selection. Less agreement exists over the origin of premating isolating mechanisms, because less difficulty exists in proposing models which account for their evolution by selection acting against inferior hybrids, as an alternative to their production in the same manner as postmating mechanisms. Many cases of "reinforcement" of premating isolating mechanisms have been reported (Levin, 1970), but it should be appreciated that the mere demonstration that selection strengthens premating isolation can in no way be regarded as compelling evidence in support of the thesis that premating isolating mechanisms evolve in this manner. This is because no satisfactory mechanism has been proposed to explain how genes, selected at a parapatric interface between two populations for their property of reinforcing premating isolation, can spread through the body

of each population outside the zone of contact. Because premating isolating mechanisms are properties of species as a whole, models of speciation must account for this fact. This is an old objection (Moore, 1957) but one which is still avoided by many advocates of the hypothesis of the selective origin of premating isolating mechanisms.

Although very brief, and not covering all suggested models for sympatric speciation, my statement did revive Moore's critical objection to the belief that speciation might occur in hybrid zones (Moore, 1957). I also drew attention once more to the groupish nature of the then prevailing way of regarding species. I believe this was the first occasion on which it was pointed out that the mere demonstration of a strengthening of "reproductive isolation" under selection does not constitute strong evidence for speciation by reinforcement. Evidence is needed to demonstrate just how any changes arising by reinforcement are stabilized in the way we can observe between natural species. Even Mayr (1963:109) accepts that isolating mechanisms are likely to decay if a population is geographically separated from its congeners: "Where no other closely related species occur, all courtship signals can 'afford' to be general, nonspecific, and variable." This is a matter that I reexamined in 1978 (chap. 3) and 1985 (chap. 12). The article was concluded with the following words:

> The studies reported on here will give some impression of the scope of Species Studies as defined above. All the studies mentioned were designed from a purely scientific point of view, but it will be noted that those involving pest animals are of considerable practical value as well. All are essentially studies of single species. In other words, they are studies at the population level of complexity. Species Studies are also an essential first stage in the study of ecosystems. This is, of course, because the organization which is apparent at the community level results from selection acting at the level of the individual within a species. The ecosystem comprises a physical environment occupied by a number of species. Ultimately its structure depends on the process of speciation although, more immediately, it may depend on the migration of species. In turn, the coadjustment of a migrant species and the ecosystem provides the selective pressure which leads to speciation under the system outlined in the introduction to this section of the paper.
>
> The significance of Species Studies might seem obvious, and yet they are rarely pursued in the way which Mayr had in mind. That much remains to be done is shown by the fact that disagreement is still widespread over so fundamental a matter as the process of speciation. We find population biologists of high standing quite unable to agree on the answer to a question such as: "Are species the product of direct selection for diversity, or are they by-products of adaptive evolution?" And yet each alternative carries with it far reaching and contrasting evolutionary implications.

Here I once more emphasized the role of speciation in ecology.

Speciation, not interspecific character displacement, was considered to be the mechanism of diversification. Competition was not evoked as an evolutionary motor, though its occurrence was not denied. Rich diversity of species reflects past circumstances that favored the divergence of isolated populations, the fixation of the changes, and their long-term persistence.

The remainder of this book records the development of many of the ideas first touched on in this small paper of 1973.

Evolution and the Recognition Concept of Species

1 The Term "Isolating Mechanisms" as a Canalizer of Evolutionary Thought

H.E.H. Paterson

I intended this paper to be the first in a series to expound the recognition concept. The next was meant to be a paper criticizing speciation by the reinforcement of isolating mechanisms. In practice the first was rejected by Nature, and the second, in an early configuration, was read at the Fifteenth International Congress of Entomology in 1976 (chap. 2) and at the Leeds meeting of the Population Genetics Group of the Genetical Society in January 1978. The latter was eventually published in 1978 (chap. 3).

It was unfortunate that the 1976 manuscript was never published, as it would have prepared the way for the 1978 paper. In this work I used the original term "mate recognition system," which was coined in Paterson (1973), but I soon realized that this was ambiguous. A second form of mate recognition is the recognition of each other as individuals by a male and female animal which have formed a long-lasting pair-bond. To avoid confusion between these alternatives, specific-mate recognition system (SMRS) was later coined.

This paper introduced in fuller and more refined form many of the ideas only touched on in my 1973 paper, mentioned in the Author's Preface, and demonstrated some of the unsatisfactory aspects of the then prevailing biological species concept.

The term "isolating mechanisms" was first introduced into the evolutionary literature by Theodosius Dobzhansky in 1935. Few terms have achieved such universal acceptance by evolutionary theoreticians of all persuasions, as can readily be established by surveying modern textbooks on evolution such as Mayr (1963, 1970), Cain (1954), Patterson and Stone (1952), Mettler and Gregg (1969), Grant (1971),

Stebbins (1971), and Murray (1972). Despite this success, I shall argue that the term is not a felicitous one and that it has had deleterious consequences for progress in the development of evolutionary theory.

Experience has shown that it is necessary to initiate all discussions such as this with a clear statement of what is meant by the term "species." Here I shall be talking about the genetical concept of species, not the taxonomic concept. Discussion is therefore restricted to sexually reproducing, biparental organisms. Essentially, the species in nature is a population of organisms which exchange genes only with other members of their own species. The gene pool of a species comprises all the genes of the individuals included in the species. Carson has focused on the essential genetical property of species by referring to "the species as a field for gene recombination" (Carson, 1957). Dobzhansky (1937a) was referring to the limits to this field for recombination when he wrote: "The expression 'isolating mechanisms' seems to be a convenient general name for all mechanisms hindering or preventing the interbreeding of racial complexes or species." Evidently Fisher (1930) had a very similar idea in his mind when he wrote that "we may infer the normal prevalence of mechanisms effective in minimizing the probability of impregnation by foreign pollen." It should be noticed that these views attribute to the isolating mechanisms the negative role of isolating the gene pool of a species from contamination by genes from other, different species' gene pools. Two more quotations will reinforce this point. In 1930 Fisher wrote: "The grossest blunder in sexual preference, which can be conceived of an animal mating, would be to mate with a species different from its own and with which the hybrids are infertile or, through the mixture of instincts and other attributes appropriate to different courses of life, at so serious a disadvantage as to leave no descendants." Dobzhansky (1937b) stated at the beginning of chapter 8, "Isolating Mechanisms": "The only way to preserve the differences between organisms is to prevent them interbreeding, to introduce isolation." The same viewpoint has constantly been reaffirmed by other authors. With clear approval, Littlejohn (1969) attributed the following to Ehrman: "Thus, the essential and primary step in the speciation process in such organisms will be the development of those devices that restrict exchange of genetic information between sympatric species (isolating mechanisms) to a level below the critical point where natural selection can no longer maintain specific distinctness." Mayr (1963) wrote: "The term 'isolating mechanisms' refers only to those properties of species populations that serve to safeguard reproductive isolation." It is this negative way of regarding species that I shall examine in this contribution. I believe it is important to do so because this viewpoint has wide implications for evolutionary theory, as I shall attempt to show. A critical examination of the concept of isolating mechanisms is perhaps timely because the views of Dobzhansky have by now almost achieved the scientifically unsatisfactory status of dogma.

In the following analysis I have been much guided by G. C. Williams' views on the analysis of adaptation (Williams, 1966).

DIFFICULTIES WITH THE TERM "ISOLATING MECHANISMS"

> A benefit can be the result of chance instead of design. The decision as to the purpose of a mechanism must be based on an examination of the machinery and on argument as to the appropriateness of the means to the end. It cannot be based on value judgments of actual or probable consequences. (G. C. Williams)

A careful look at the term "isolating mechanisms" leads me inevitably to the conclusion that Dobzhansky believed that they are adaptations serving to *isolate* the gene pool of a species. In other words, that they have been perfected by natural selection to that end. The use of the word "mechanism" implies to me "that the machinery involved was fashioned by selection for the goal attributed to it." If this is accepted, a further implication is that natural selection has a direct role in the production of species diversity.

This last conclusion at once makes the term unsatisfactory, for the following reasons. First, compelling evidence exists only for speciation in total allopatry. If this statement appears to be too strong, consideration should be given to identifying at least one case where species *must* have come into existence by the reinforcement model of speciation, the one model which involves the direct action of natural selection for species diversity.

Another reason for doubting that isolating mechanisms are always the product of natural selection, as the name seems to imply, is that it is well established that the so-called postmating isolating mechanisms (hybrid inviability, hybrid sterility, and hybrid breakdown, in Dobzhansky's 1970 terminology) cannot have been produced under natural selection (see, for example, Mecham, 1961).

These two points show clearly that in many cases isolating mechanisms have come into existence by "chance" rather than "design." In other words, their role in isolating a species' gene pool is an incidental one, and not the one for which they were selected.

It has been shown above that the use of the term "isolating mechanisms" carries with it the necessity to accept speciation by reinforcement as the way in which they evolve. Thus, it would be difficult to sustain the view that isolating mechanisms are adaptations shaped by natural selection for the role of isolating a species' gene pool, if the reinforcement model of speciation ultimately proves unsatisfactory.

In 1957 John Moore gathered together a number of difficulties, mostly of a population genetic nature, which required detailed attention from advocates of the model of speciation by reinforcement. The subject was raised and studied again by Crosby (1970) using a computer model to simulate a reinforcement situation in plants which were stipulated to be entomophilous hermaphrodites which bred at random. The model was arranged rather artificially to provide selection for a temporal isolating mechanism (flowering time). During the preliminary runs Crosby rediscovered the properties of the unstable polymorphism based on heterozygote disadvantage. This he artificially overcame by the constant addition of plants of appropriate genotype. The simplifying assumptions made the model unconvincing even for the simple plant situation studied, let alone for animals with separate sexes. I do not believe Crosby's study furthered the cause of speciation by reinforcement.

Mayr (1963, 1970) mentions Moore's paper but scarcely does justice to it, though he does judge it as "able." Dobzhansky (1970) does not mention it. The most extensive coverage of the subject is by Murray (1972). He concluded that there is "a considerable body of evidence favoring the occurrence of selection for the improvement of isolating mechanisms, once incipient isolation has been established." I find this conclusion debatable in view of Moore's criticisms, and for other reasons, and shall examine the reinforcement model critically in another paper. Suffice it to say here that there is a vast difference between selection favoring "the improvement of isolating mechanisms" and the evolution of species-specific isolating mechanisms as a direct consequence of selection for positive assortative mating in a zone of secondary overlap. The latter alternative has not yet been proved to occur.

In 1970 Dobzhansky provided the following list of isolating mechanisms, which differed from the one he gave in 1951 by not including geographic isolation. Even so, the list includes a great diversity of phenomena.

1. Premating or prezygotic mechanisms
   a. Ecological or habitat isolation
   b. Seasonal or temporal isolation
   c. Sexual or ethological isolation
   d. Mechanical isolation
   e. Isolation by different pollinators
   f. Gametic isolation

2. Postmating or zygotic isolating mechanisms
    g. Hybrid inviability
    h. Hybrid sterility
    i. Hybrid breakdown

Another disadvantage of the term "isolating mechanisms" is that this name misleadingly brings under one heading, in an artificial and unrealistic way, the above array of very distinct phenomena. I have already shown that the model of speciation by reinforcement cannot account for the evolution of, at least, a number of these. All of them are biologically meaningful, and each deserves study in its own right. What is necessary is an assessment of the true function each was evolved to serve. Grouping them and naming them in this way is tantamount to obscuring their true significance.

AN ALTERNATIVE VIEW OF THE GENETICAL SPECIES

Dobzhansky (1937a) realized that "a thorough understanding of the nature and functioning of isolating mechanisms is essential, because without it no trustworthy picture of the mechanism of evolution can be drawn." It is for this reason that I have examined isolating mechanisms rather critically, though in broad terms. I now wish to advance an alternative view of genetical species which seems to me to offer a more acceptable interpretation of what determines the limits of the species' gene pool.

The essential idea in this approach is that natural selection is not involved in the negative role of preventing hybridization between species, but is strongly involved in the positive role of maintaining mate recognition. This is not a particularly new idea, and I am sure that to less sophisticated people than modern biologists the pairing and breeding of species "each after its own kind" would be rather to be expected. I am not suggesting that evolutionary theorists are not aware of this idea, but I am saying that I am not aware of one who follows this positive approach consistently. After reading a clear exposition of the positive approach, one frequently reads on to find a relapse into the negative view. Examples of this ambivalence can be found throughout Mayr (1963); pages 20–21 and page 91 can be cited as particular examples. Even Muller (1942) and Moore (1957) continued speaking of "isolating mechanisms."

The precise recognition of conspecific mates thus ensures that gene exchange is restricted to within the species population, which is then the "field for gene recombination." This recognition is called "species recognition" by Mayr (1963:95), but I prefer the term "mate recognition" as a more precise statement of what is involved. The mate recognition system has often been discussed with emphasis placed on one aspect of the system or another according to the outlook of the author. Figure 1 provides a general scheme which emphasizes

1. that the system is a stimulus-response-reaction chain;
2. that each step in the chain is under genetic control;
3. that each stimulus, response, and reaction is characterized by an optimal value with a characteristic variance; and
4. that the successive steps in the chain are coadapted and the whole system is subject to stabilizing selection.

The mate recognition system of a species includes the species-specific courtship pattern, but may include a number of other component systems, such as the cellular recognition between egg and sperm or pollen and stigma. It seems likely that even the very first sexually reproducing eukaryote was possessed of a mate recognition system, albeit a simple one. It possibly comprised a simple chemical signaling system involving specific receptor sites. The meeting of potential sexual partners could have relied on chance collision meetings. For fertilization to occur efficiently in more advanced, motile organisms, such as among the metazoa, a more reliable method for bringing the sexual partners together is a necessity. This imperative led to the further elaboration of the courtship pattern to include a component which functions to attract conspecific potential partners from a distance. Yet another need arose with the evolution of long-lived organisms, the need to synchronize the physiological condition of the partners for fertilization.

This does not exhaust the components of sexual behavior, but it is sufficient to show that courtship behavior, and the mate recognition system as a whole, comprise several components, each serving a distinct function and therefore having arisen under a different selective regime, a vital point to grasp if an understanding of the genetical basis of the speciation process is to be aimed for.

DISCUSSION

In 1966 G. C. Williams discussed "adaptation," "function," "mechanism," and other related terms, urging restraint and care in their use. These strictures have been kept in mind while writing this discussion of one of the most influential terms in evolutionary genetics.

The term "isolating mechanisms" has been shown above to be inappropriate and misleading. Inappropriate, because the name inevitably implies that the phenomena listed by Dobzhansky (1937a, 1951, 1970) and other authors were shaped by natural selection for the

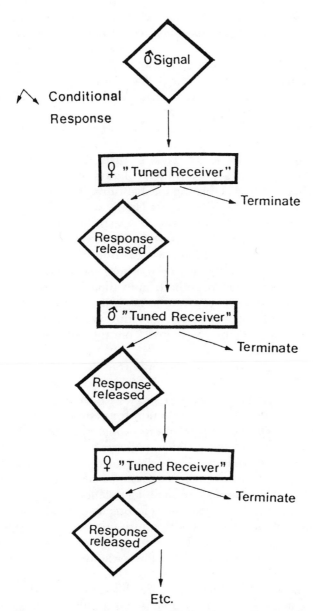

Figure 1. Diagrammatic representation of the sequence of events involved in a mate recognition system of a biparental species of animal. The number of stages in the sequence is characteristic of a species.

function of *isolating* one species' gene pool from all others. It was demonstrated that many of the listed "isolating mechanisms," the "postmating isolating mechanisms," could not have evolved to serve this function. In order to understand the evolution of a species characteristic, whether structural, behavioral, or physiological, its essential function must be identifiable. As Williams has emphasized, a character shaped by selection to serve a particular function often has other, incidental (pleiotropic) *effects*.

The term "isolating mechanisms" is misleading because it canalizes thought into particular channels with the result that "effects" are misconstrued as "functions." This, in turn, deviates research away from the true selective basis of the evolution of a particular species-specific characteristic ("isolating mechanism"). This has had insidious results. Much effort has been devoted to the study of "reinforcement," with very meager returns by way of convincing evidence. How all-pervading is this influence is shown by the frequency with which one finds statements such as the following: "Both pheromone strains [of moth] occur sympatrically near Harrisburg, Pennsylvania. . . . If hybrids were less fit than pure bred individuals, we would expect reproductive isolation to be strongest in such an area, because insects that would participate in interstrain matings would be selected against by having their progeny eliminated from the population" (Liebherr and Roelofs, 1975).

Despite long lists of papers in support of the reinforcement model (Levin, 1970), I should be surprised if it were possible to identify with certainty a single pair of related species which *must of necessity* have evolved as a consequence of selection for reproductive isolation. In fact, the same can be said of all the proposed models of speciation other than the model under which speciation occurs as a consequence of total extrinsic isolation. The prima facie evidence is strong, for example, that *Pinaroloxias inornata* (Geospizinae, Fringillidae) evolved on Cocos Island to species status in total isolation from all other members of its family. To invoke the reinforcement model of speciation to account for such cases would require postulating the existence of a former congener, now extinct, in the absence of any evidence for its existence. Such an argument to me is wholly unconvincing though it was used by Wallace in 1889 and by Dobzhansky.

The Cocos Island finch, and cases like it, do provide a clue to a more acceptable way of viewing genetical species. All biparental species are faced with the problem of potential sexual partners recognizing each other, and all genetical species must, one would predict, be equipped to do this. Specific-mate recognition is achieved by the "mate recognition system," comprising a sequence of steps involving the emission of a signal, its reception by the potential partner, its processing by the central nervous system which conditionally releases a response to be similarly received, processed, and responded to by the first partner, etc. Such a system is very constant, being under strong stabilizing selection because the stages are coadapted. It is obvious that such a system did not escape the notice of Dobzhansky and others. But they classified mate recognition systems as ethological or behavioral or sexual isolating mechanisms, thus elevating the *effect,* isolation, to the status of the *function,* and consequently obscured the basis of its evolution. By recognizing the system's true function, mate recognition, we are in a better position to understand or elucidate its evolution. Furthermore, we are then in a position to recognize most of the other so-called isolating mechanisms as being incidental effects determined by genes selected originally to serve other functions. Positive assortative mating, the consequence of the mate recognition system, ensures that the species is the field for gene recombination.

Looked at in this way the Cocos Island finch, the African hammerhead (*Scopus umbretta*), the Madagascan partridge (*Margaroperdix madagascariensis*) (Frost, 1975), and other very distinct species which seem to have arisen in total isolation from any close relative offer no special problems of interpretation.

The most important consequence of accepting this positive viewpoint is that it brings into clear focus what should be looked at in speciation studies. Instead of looking for signs of reinforcement and other forms of character displacement, we should be assessing the consequences and details of allopatric divergence leading to speciation and deciding how effectively these can explain adaptive radiation, and the evolution of communities, ecosystems, and faunas without extensive resort being made to invoking the dubious concept of character displacement in its various guises.

REFERENCES

Cain, A. J. 1954. *Animal Species and Their Evolution.* London: Hutchinson.

Carson, H. L. 1957. The Species as a Field for Gene Recombination. In *The Species Problem,* ed. E. Mayr, 23–38. Publication No. 50. Washington, D.C.: American Association for the Advancement of Science.

Crosby, J. L. 1970. The evolution of genetic discontinuity: Computer models of the selection of barriers to interbreeding between species. *Heredity* 25:253–97.

Dobzhansky, T. 1935. A critique of the species concept in biology. *Philosophy of Science* 2:344–55.

———. 1937a. Genetic nature of species differences. *American Naturalist* 71:404–20.

————. 1937b. *Genetics and the Origin of Species.* New York: Columbia University Press.

————. 1951. *Genetics and the Origin of Species.* 3d ed. New York: Columbia University Press.

————. 1970. *Genetics of the Evolutionary Process.* New York: Columbia University Press.

Fisher, R. A. 1930. *The Genetical Theory of Natural Selection.* Oxford: Oxford University Press.

Frost, P.G.H. 1975. The systematic position of the Madagascan partridge *Margaroperdix madagascariensis* (Scopoli). *Bull. B. O. C.* 95:64–68.

Grant, V. 1971. *Plant Speciation.* New York: Columbia University Press.

Levin, D. A. 1970. Reinforcement of reproductive isolation: Plants versus animals. *American Naturalist* 104:571–81.

Liebherr, J., and W. L. Roelofs. 1975. Laboratory hybridization and mating period studies using two pheromone strains of *Ostrinia nubilalis*. *Ann. Entomol. Soc. Am.* 68:305–9.

Littlejohn, M. J. 1969. The Systematic Significance of Isolating Mechanisms. In *Systematic Biology*, ed. C. G. Sibley, 459–82. Washington, D.C.: National Academy of Sciences.

Mayr, E. 1963. *Animal Species and Evolution.* Cambridge, Mass.: Belknap Press of Harvard University Press.

————. 1970. *Populations, Species and Evolution.* Cambridge, Mass.: Belknap Press of Harvard University Press.

Mecham, J. S. 1961. Isolating Mechanisms in Anuran Amphibians. In *Vertebrate Speciation*, ed. W. F. Blair, 24–61. Austin: University of Texas Press.

Mettler, L. E., and T. G. Gregg. 1969. *Population Genetics and Evolution.* Englewood Cliffs, N.J.: Prentice-Hall.

Moore, J. A. 1957. An Embryologist's View of the Species Concept. In *The Species Problem*, ed. E. Mayr, 325–38. Publication No. 50. Washington, D.C.: American Association for the Advancement of Science.

Muller, H. J. 1942. Isolating mechanisms, evolution and temperature. *Biological Symposia* 6:71–125.

Murray, J. 1972. *Genetic Diversity and Natural Selection.* Edinburgh: Oliver and Boyd.

Patterson, J. T., and W. S. Stone. 1952. *Evolution in the Genus Drosophila.* London: Macmillan.

Stebbins, G. L. 1971. *Processes of Organic Evolution.* Englewood Cliffs, N.J.: Prentice-Hall.

Wallace, A. R. 1889. *Darwinism.* London: Macmillan.

Williams, G. C. 1966. *Adaptation and Natural Selection.* Princeton, N.J.: Princeton University Press.

2 The Role of Postmating Isolation in Evolution

H.E.H. Paterson

On August 23, 1976, in Washington, D.C., at the Fifteenth International Congress of Entomology, a Symposium on the Application of Genetics to the Analyses of Species Differences was chaired by Marshall Wheeler. The contributors were Francisco Ayala, Michael White, Esa Suomalainen, and I. Guy Bush's role was to review and summarize. Suomalainen was not able to attend, and his paper was read for him.

Fortuitously, Ayala read a shortened version of his paper of 1974 from Evolution, *which was convenient because the audience was thereby reminded of the sort of argument that I was seeking to undermine. Although Muller and Moore had earlier criticized the reinforcement model of speciation, both continued to speak of "isolating mechanisms" and reproductive isolation. In this talk I criticized the isolation concept and recommended that it be abandoned in favor of the recognition concept. For most evolutionists this was the first time they had heard these views, and their response was largely unsympathetic. Only Muller's intellectual descendants liked it.*

I also emphasized the dependence of much of ecological theory on the nature of species and speciation, as I did in the 1973 article, and called into question the evolutionary significance of "sterility" in its various guises. This was further developed in later papers, particularly those of 1981 and 1988 (chaps. 5 and 14).

Subsequently a version of this paper was circulated to a number of colleagues. Templeton (1979) referred to this manuscript, and this stimulated me to write my 1980 note (chap. 4).

It should be appreciated that the paper is written in a style appropriate for an oral presentation. References have been interpolated, and a reference section has been added.

11

After more than one hundred years of evolutionary theory there still exists compelling evidence for only one mode of speciation: speciation in total allopatry.[1] Yet many authors appear reluctant to accept this fact. In their writings the model which is most often given prominence is the one attributed to A. R. Wallace (1889), speciation by reinforcement of incipient reproductive isolation. This preference for the Wallace scheme has worried me because it entails difficulties which appear to demand more serious consideration than has been forthcoming. I hope that this is sufficient reason for raising the subject once more. It is important, I think, that the two principal speciation models should receive constant review and not be allowed to suffer the fate of becoming dogma. This is because each has quite distinct implications which bear strongly on evolutionary ideas in general and ecological theory in particular. In theorizing we need to build on firm foundations so that our edifices are not "white-anted" from below.

The fundamental difference between the two models stems from the fact that in the Wallace model there is an essential role for natural selection in the production of species diversity on earth, whereas, under the allopatric model, species diversity is not selected for at all: it is a by-product of adaptive evolution in allopatry. This difference is of great significance for our thinking about the evolution of ecosystems, communities, and faunas, and in other ways.

To avoid misunderstanding, let me state clearly my attitude to the concept of species before proceeding. Throughout this discussion I shall be concerned with the genetical concept. By this I have in mind the same concept as had Carson (1957) when he coined his memorable phrase, "the species as a field for gene recombination." The biological limits to this field are set by what I call the "mate recognition system," which has the following components, repeated, in some cases, several times (see Fig. 1, chap. 1):

1. A specific signal with releasing properties is emitted by the initiating partner.
2. This is received by a complementary sensory system specific to the second partner.
3. The message is filtered by a coded component of the central nervous system, which conditionally releases a response.
4. This is received by a complementary sensory system specific to the first partner.
5. The message is filtered by a coded component of the central nervous system which conditionally releases a response, etc.

It seems certain that all these stages are largely genetically controlled. Equally certain it is that succeeding stages are closely coadapted, which, in turn, implies that the whole system is under stabilizing selection.

The very first primitive, sexually reproducing eukaryote must have possessed a mate recognition system, albeit a simple one. It probably comprised simple chemical signals received by specific receptor sites.

For syngamy to occur efficiently, especially in the Metazoa, chance collision meetings between potential sexual partners would need to be improved upon, though they may suffice in some simple organisms. This imperative led to the further evolution of the courtship system so as to include a component which functions to attract mates from a distance.

With the evolution of long-lived organisms, yet another need developed: the need to synchronize the physiological condition of the partners for fertilization. This, then, is a bare outline of the evolution of complex courtship patterns.

In higher Metazoa it is possible to detect behavioral components of the courtship system which facilitate the meeting of the sexes and the achieving of fertilization. These include restriction of organisms to specific habitats, meeting of the sexes on food, meeting on hosts, etc. Also relevant are the daily activity rhythms. Synchrony in physiological state is also achieved in diverse ways. These components are superimposed on the true mate recognition systems and should be recognized [later recognized as a broader "fertilization system" (chap. 12)].

These views may seem conventional and generally acceptable, but, nevertheless, their implications do not always seem to have been fully appreciated, in particular, the fact that such a coadapted system is not readily disturbed in the organism's habitat by mutation or selection. This point is of prime importance in assessing the value of any model of speciation. Consider, too, its significance for the theory of sexual selection.

Let us now return to an analysis of the reinforcement model. Wallace's model of speciation involves several stages, which can be stated as follows:

1. The budding off of a daughter population from the main body of a species, followed by its virtually complete isolation by extrinsic barriers.
2. The genetic divergence of the daughter population from the parent population because of its adaptive evolution in response to a distinct environment.
3. The meeting of the genetically diverged populations. If the interpopulation hybrids are selectively disadvantageous, natu-

ral selection will act to favor genes which enhance positive assortative mating. This may lead to the differentiation of two distinct genetical species.

(Note that it postulates that any postmating isolation arises as a by-product of adaptive evolution.)

I believe that we tend to accept this reasonable-sounding model too uncritically. I feel we need to look at it in detail in order to test its credibility.

Under this model we are assigning a major role to selection in the evolution of positive assortative mating. It seems to me fair, therefore, to assume that no progress toward positive assortative mating has occurred in allopatry. If this is acceptable, we can analyze the situation simply by applying the standard algebraic treatment of the unstable polymorphism due to heterozygote disadvantage.

The results of such an analysis can be readily appreciated from the following figure [see Figs. 1 and 2 in chap. 3]. In the figure the mean fitnesses of the two pure parental populations are set at 1 for simplicity. It is, of course, important to know the exact circumstances relating to the meeting of the two populations, but in principle it is clear that there will be a rapid decline in the rarer allele (or super-allele).[2]

A question which never seems to be considered is whether there is sufficient time for reinforcement to occur according to the Wallace model. It seems to me important to examine the interactions of time and the selective values in this situation.

Another problem with this aspect of the model is that selection in favor of reinforcement will be opposed by the strong selection stabilizing the mate recognition system.

Even more difficulties call for answers. If two diverged populations should meet again along an extensive front, or at more than one place, there seems to be no reason to believe that the same pattern of reinforcement should evolve at more than one point. Evolution is, after all, essentially opportunistic, acting on the most suitable phenotype available. Yet, observations of variation in mate recognition systems suggest, rather, that each species is essentially homogeneous in this respect.

Dobzhansky (1940:389) (a strong supporter of the Wallace model) himself raised another important problem which has been looked at by Moore (1957) and, more recently, by Crosby (1970). This relates to the problem of fixing alleles, which are selectively advantageous at the interface, in a population as a whole.

Such alleles, of course, are the alleles selected to reinforce positive assortative mating. Dobzhansky's early answer to this puzzle was: "If certain genes are favorable in a part of the species area and neutral elsewhere, they will eventually diffuse throughout the species by migration."

To me this explanation is not satisfactory for the following reasons. Any gene selected at the interface of the two populations to reinforce assortative mating must, by definition, affect the mate recognition system and therefore be opposed by the stabilizing selection to which this system is subject.

If it spreads at the interface of the populations it must be because selection against the heterozygotes outweighs the stabilizing selection on the mate recognition system.

However, the new allele for assortative mating will not diffuse into the body of the population as Dobzhansky expected, because, away from the interfaces it will be opposed by the stabilizing selection due to coadaptation which maintains the stability of the mate recognition system.

I believe these are serious difficulties with the Wallace model, and there are others. Yet, many authors seem reluctant to invoke the wholly allopatric model, as I have already mentioned. Cases are known when formerly allopatric populations have been observed to come together and behave as distinct genetical species. Rather than accepting cases like this as the result of wholly allopatric speciation, we find rather procrustean explanations being advanced to bring them under the heading of the Wallace model.

The following is an example from a leading theorist, Theodosius Dobzhansky (1940:389), who, following A. R. Wallace (1889) in *Darwinism*, is attempting to account for the reproductive isolation observed between a recent invader and a local species: "It seems reasonable to suppose that immigration had occurred repeatedly, and that migrants have become established only after the development of physiological isolation due to previous intrusions."

Unfortunately, cases are known where multiple invasions of this sort are incredibly unlikely, and yet the populations behave as good species.

SUMMARY

It is evident that two populations will not be able to coexist if hybrid sterility is the only protection existing for their gene pools. The gene pools *are* effectively protected, but the rarer form will be rapidly eliminated.

If the reinforcement model can be validated, and the difficulties raised above satisfactorily accounted for, then hybrid sterility or disadvantage would possess a major role in evolution, that of providing the basis for reinforcement to occur.

If, however, this speciation model ultimately proves unsatisfactory, only minor evolutionary significance will attach to hybrid disadvantages. They will have some role in determining the fate of related diverged populations which share a common mate recognition system. In plants a wider role is perhaps possible.

To complete my talk I should like to discuss an application of the type of thinking I have just been indulging in to the solution of a puzzling old problem that has been with us since the thirties and which has caused the writers of general evolutionary texts much perplexity: this is the *Culex pipiens* complex in mosquitoes. I have been looking at it for many years, but since 1967 I have been able to study four sympatric members in Western Australia.

In 1970 I was joined by S. J. Miles, to whom was assigned the role of examining the complex without reference to cytoplasmic incompatibility.

Workers in Europe had found cytoplasmic incompatibility in making crosses between various laboratory populations of *C. molestus* and also between cage populations of *C. quinquefasciatus*. This incompatibility, which sometimes occurs in both directions, but more often in one only, has been shown by Laven (1953) to have a cytoplasmic inheritance pattern. Making the assumption that this postmating phenomenon is not connected with speciation, but is an intraspecific event, R. Irving-Bell was asked to investigate it. The results of this investigation have been heard at this conference.

Briefly, Miles was able to use genetical, zoogeographical, and behavioral evidence to show that the complex can readily be clarified and that it comprises six or seven old-world species (Table 1).

Irving-Bell (1974) and, independently, Yen and Barr (1971), have shown that the cytoplasmic factors are probably the rickettsia *Wohlbachia pipientis*. Furthermore, she has shown that, in certain of the species delineated by Miles, there are no rickettsia symbionts present in the gonads. She has also provided good evidence to indicate that many of the mating types recognized by other workers have evolved in the laboratory. This means that Laven's maps of Europe (Fig. 1) showing so-called distribution patterns of mating types are at least partly suspect. In fact, good evidence for distinct mating types within species of this complex in nature is rare.

In 1959, and on a number of subsequent occasions, Laven has dealt with the implications of his work for speciation theory. He has outlined a method of so-called speciation which reaches completion

Table 1. Culex pipiens complex

Species	Origin
Culex pipiens[a]	Europe, North America
Culex molestus[a]	Europe, North America
Culex quinquefasciatus[a]	Oriental region
Culex pallens[a]	Eastern palearctic
Culex australicus	Australia
Culex globocoxitus	Australia
African "*Culex pipiens*"	Africa

[a]Rickettsia present.

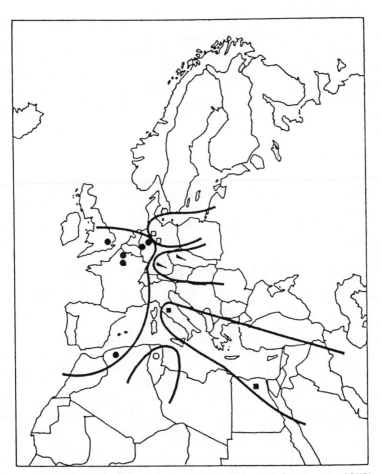

Figure 1. Distribution of "crossing types" in Europe (after Laven, 1959).

when two populations have evolved which have their respective gene pools protected by a complete sterility barrier. He refers to such a sterility barrier as the "perfect isolating mechanism."

As already stated, the complete sterility barrier does perfectly protect the integrity of the two gene pools, but, unfortunately, this is of only academic interest because it also prevents the two forms from coexisting. After all, mere geographic separation achieves as much!

I believe, therefore, that the reason why earlier workers, including many very distinguished ones, have failed to elucidate the complexities of the *Culex pipiens* complex is that the possible evolutionary significance of postmating isolation was misjudged. They believed it was related to the specific divergence of populations, whereas it is an intraspecific phenomenon.

There are other cases, such as the *Drosophila paulistorum* complex, which I should like to see reanalyzed in the manner we have used on the *Culex pipiens* complex, that is, by playing down the role of the postmating isolation and placing emphasis on the mate recognition systems. (This is because, in this species complex, cytoplasmic incompatibility is determined by symbiotic prokaryotes, and cases of change in mating type have occurred in the laboratory.)

REFERENCES

Carson, H. L. 1957. The Species as a Field for Gene Recombination. In *The Species Problem*, ed. E. Mayr, 23–38. Publication No. 50. Washington, D.C.: American Association for the Advancement of Science.

Crosby, J. L. 1970. The evolution of genetic discontinuity: Computer models of the selection of barriers to interbreeding between subspecies. *Heredity* 25:253–97.

Dobzhansky, T. 1940. Speciation as a stage in evolutionary divergence. *American Naturalist* 74:312–21.

Irving-Bell, R. J. 1974. Cytoplasmic factors in the gonads of *Culex pipiens* complex mosquitoes. *Life Sciences* 14:1149–51.

Laven, H. 1953. Reziprok Unterschiedliche Kreuzbarkeit von Stechmucken (Culicidae), und ihre Deutung als plasmatische Vererbung. *Zeitschrift für Vererbungslehre* 85:118–36.

———. 1959. Speciation by cytoplasmic isolation in the *Culex pipiens* complex. *Cold Spring Harbor Symposia of Quantitative Biology* 24:166–73.

Moore, J. A. 1957. An Embryologist's View of the Species Concept. In *The Species Problem*, ed. E. Mayr, 325–38. Publication No. 50. Washington, D.C.: American Association for the Advancement of Science.

Wallace, A. R. 1889. *Darwinism*. London: Macmillan.

Yen, J. H., and A. R. Barr. 1971. New hypothesis of the cause of cytoplasmic incompatibility in *Culex pipiens* L. *Nature* 232:657–58.

NOTES

1. This apparently provocative statement will require qualification only when some way is found to demonstrate that a particular pair of species *must* have arisen by another process.
2. When $s = 1$, it is the rarer form (population) which declines; when $s < 1$, it is the rarer allele at the locus for incompatibility that declines.

3 More Evidence against Speciation by Reinforcement

Paterson, H.E.H. 1978. More evidence against speciation by reinforcement. *South African Journal of Science* 74:369–71.

The ideas in this paper had been tested in January 1978 at Leeds before a professional audience at the annual meeting of the Population Genetics Group of the Genetical Society.

It is compactly written because of editorial constraints on the length of papers. It is informative on the recognition concept, though in a rather nonconfrontational way that did not force readers to consider key points. My main objective was to criticize the idea of reinforcement, which was espoused particularly by Dobzhansky, but also by Ernst Mayr (e.g., 1963:551, paragraphs 3 and 4). Earlier, Muller and Moore had effectively criticized speciation by reinforcement, and Ernst Mayr was, in practice, ambivalent, opposing it at times and yet, at others, advocating it. Nevertheless, these three workers retained their allegiance to the isolation concept and referred to "isolating mechanisms," "reproductive isolation," and so on. I emphasized the last scientific note by Jasper Loftus-Hills, who had worked for his Ph.D. degree at the University of Melbourne with Murray Littlejohn. Because of his early work it is particularly significant that he should have come to the conclusion that there existed then little evidence in support of reinforcement.

This paper was most significant because it was the first to criticize all sources of support for the reinforcement theory. I am not aware of any other worker having done this as comprehensively. Points made in this brief paper have been amplified in my 1982 paper (chap. 8) and others.

An important insight which is still not fully appreciated by many was my seeing that the laboratory experiments which purported to support the idea of reinforcement by selection were all flawed. This is because, in the design of the experiments, the two interacting populations were always kept equal in size. This is unrealistic, because we are talking about a metastable situation as an examination of what the population genetical model reveals. The paper in this volume by Lambert, Centner, and Paterson (1984) (chap. 11) models this situation, and so do papers by Spencer, McArdle, and Lambert (1986) and Spencer, Lambert, and McArdle (1987).

In this paper I mentioned some work by Hugh Robertson which demonstrated the unreality of the experimental design which till then had always been adopted. Later, this work was repeated and reported on in full detail by Harper and Lambert (1983), who fully confirmed Robertson's preliminary study.

It is indeed curious that authors, from Fisher onward, have always believed that natural selection must act against hybrids by bringing about reinforcement. Why, one wonders, does natural selection not take the more straightforward and obvious course and eliminate the cause of the hybrid disadvantage? One suspects the power that preconception exerts on interpretation.

This was the first place where I emphasized the stability of the specific-mate recognition system because of its coadapted nature. This has been followed up with some excellent studies by Henderson and Lambert (1982), which much extended the earlier work I had cited in 1978.

I tried citing Petersen (1905) instead of my own views to reduce impatient rejection of the ideas in this paper, and I used a quotation from Crosby in an attempt at changing the conventional viewpoint in an unobtrusive way, but with no perceptible effect.

It is a fact that each species of animals has devices which permit the recognition and the bringing together of conspecific individuals of opposite sex with such a degree of certainty that hybridization occurs only as an abnormal exception. (W. Petersen, 1905)

Petersen's perceptive words capture the essence of what we understand as species in nature. This can be stated in another way: *members of a species share a common specific-mate recognition system.* Pair formation in sexually reproducing organisms follows positive acts of recognition.[1,2] Thus, a new species arises when all members of a subpopulation of an existing species acquire a new specific-mate recognition system (SMRS). Exactly how this occurs in nature is a matter of debate, and a diligent search through the literature reveals a confusing number of suggested models.[3] In searching for perspective in this field it should be observed that only two models command widespread support. These are speciation in allopatry[4] and speciation by reinforcement.[5] It should also be noted that there is strong prima facie evidence for only one of the many models, and that is the allopatric model. I am not aware of any evidence which compels acceptance of any other model, including the reinforcement model. What evidence is available is very indirect.

The aim of this article is to add to the earlier criticisms of the reinforcement model.[6,7] I take the trouble to do this because this model features so importantly in current theoretical and applied biological thinking.[8,9] I believe that it is favored in this way because, in contrast to the allopatric model, it provides a direct role for natural selection in the production of new species, and, hence, in the production of species diversity. According to the allopatric model, new species arise as the incidental consequence of differential adaptive evolution in isolated subpopulations of an original parental species.[7] Although cogent, the earlier criticisms have not covered all the points of weakness in the reinforcement model, and they have made curiously little impact.

A CONTEMPORARY FORMULATION OF
THE REINFORCEMENT MODEL

To avoid introducing any bias, a recent statement of the reinforcement model by Ayala et al.[10] will be examined:

> In sexually reproducing organisms speciation most generally occurs according to the model of "geographic speciation." Two main stages may be recognized in this process. First, allopatric populations of the same species become genetically differentiated, mostly as a consequence of their adaptation to different environments. This genetic differentiation can only occur if the populations are geographically separated for some time, and there is no, or very little, migration between them. The second stage takes place when genetically differentiated populations come into geographic contact. If the gene pools of two populations are sufficiently different, progenies from interpopulational crosses are likely to have less fitness than progenies from intrapopulational crosses. Mating preferences are affected by genes. Alleles that decrease the probability of mating with individuals of a different population will, then, be selected [for], while alleles increasing the probability of intrapopulational mating will be favored by natural selection. Eventually, the process may result in two reproductively isolated populations, and thus two different species.

Ayala et al. cite two *Drosophila* subspecies pairs as examples of allopatric populations which have diverged genetically and which are now ready to be subjected to selection for increased positive assortative mating should they become sympatric: *D. w. willistoni* and *D. w. quechua*; and *D. e. equinoxialis* and *D. e. caribbensis*.

AN ALTERNATIVE APPROACH

It should be noticed that the above statement of the reinforcement model provides no intimation that an alternative evolutionary response to the meeting of the populations is possible. In fact, if the problem is examined in the light of deterministic population genetical theory, a quite different, and much simpler, outcome will be expected.[11,12] This can readily be demonstrated as follows:

For conceptual clarity let it be supposed that the daughter population has diverged partly due to the fixation of an autosomal chromosome translocation. Accordingly, interpopulational hybrids will be more or less infertile as a consequence of the well-known properties of translocation heterozygotes. The populations will thus possess properties quite comparable with those cited by Ayala et al. As in these, individual organisms will mate at random in the area of sympatry, regardless of their karyotype. Such a situation can legitimately be analyzed using the standard algebraic model for heterozygote disadvantage. If, for example, the two types of chromosomal homozygote are assumed to be equally fit, a metastable equilibrium point will occur at $q = 0.5$. Why the equilibrium is metastable is clear from Figure 1, which shows that any displacement of q from the equilibrium point will lead to the rapid extinction of the rarer chromosome arrangement. In Figure 2 the progress to extinction is illustrated under three distinct selective pressures.

It should be noticed that if the heterozygotes are completely sterile or inviable ($s = 1$), the rarer *population* will become extinct, but when the heterozygotes are somewhat fitter ($1 > s > 0$), it is the rarer *chromosome arrangement* that is eliminated by selection. In this case genes on other linkage groups, and on the more abundant chromosome arrangement, remain in the population gene pool. [p. 370]

EVIDENCE FROM THE LITERATURE

When two formerly allopatric populations become sympatric after diverging genetically, there are, thus, two possible nontrivial outcomes, not just one. Few authors appear to have noted this.[13-17] It would appear certain that natural selection will in all cases act to

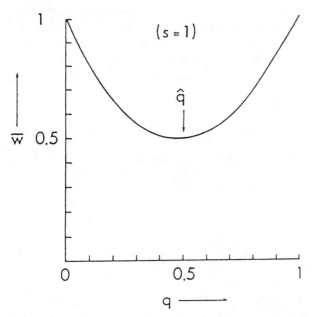

Figure 1. Population mean fitness ($\overline{w}$) related to relative gene frequency (q) assuming a coefficient of selection of unity and heterozygote disadvantage ($1 > 1 - s < 1$).[11]

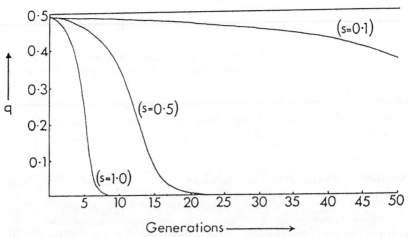

Figure 2. Decline in gene frequency (q) with time for various coefficients of selection and heterozygote disadvantage.[11] Equilibrium is at $q = 0.5$.

increase mean fitness by eliminating the genetic basis of the heterozygote disadvantage.

A study by Lewis[18] of two closely related species of *Clarkia*, which are intersterile but which are not distinguished between by insect pollinators, supports this view. At three of the four study sites where the species were planted together, the rarer species went to extinction within the study period of five generations. No finality had been achieved at the fourth site when the experiment ended.

A somewhat comparable study by Mettler[19] involved population cage experiments with mixed populations of *Drosophila arizonensis* and *D. mojavensis*. In these studies the *arizonensis* X-chromosome was either grossly reduced in frequency or eliminated by selection. This identified the X-chromosome as a disadvantageous element in the hybrid gene pool. A recent cytological study of the grasshopper *Caledia captiva* is particularly informative[17] as it provides evidence of such a process occurring in the wild.

Two computer simulation studies have been made of reinforcement models not unlike that described by Ayala et al. The first is part of Crosby's study.[15] When this simulation was run the rarer population was rapidly eliminated, as would be expected from the negative heterosis predictions, and Crosby was obliged to boost the numbers of the rarer population in later runs of the program.

The second computer study has been outlined by Wilson.[13] It, too, confirmed the predictions of the deterministic model, but, for reasons which were not made explicit, it was anticipated that reinforcement would also occur. Wilson visualized the final act of speciation as a race between these two processes and concluded that under limited circumstances reinforcement might sometimes win. However, the published details are insufficient to dispel suspicions that the author did not take into account the coadaptation of male and female parts of the SMRS. If this is so, the likelihood of a win for reinforcement is further diminished.

There have been numerous experiments designed to simulate the situation outlined by Ayala et al. Six of these have been examined.[20-25] All reported some success in attaining improved positive assortative mating, though in no case was complete isolation achieved. Attention to detail has, however, revealed that all show a fundamental flaw in design. In all six cases the numbers of the two populations were either maintained at parity in each generation, or they were adjusted to prevent the rarer population being eliminated. This action at once vitiates the experiments as simulations of the reinforcement model and demonstrates that no falsification of the negative heterosis predictions has, in fact, been achieved. This analysis is supported by an unpublished study by Hugh Robertson in which he repeated Crossley's experiment[25] but allowed the population which became rarer to decline without interference. Under these circumstances the population marked with *vestigial* was eliminated in eight generations.

Another deficiency with these experiments is that in none was it demonstrated that the genetic basis of the observed shift had become fixed in the populations. Yet such a demonstration would surely be necessary if the simulation of speciation is to be achieved in a credible manner. I say this because stability of specific-mate recognition systems is a striking feature of species in nature and is less variable geographically than are other characters which are not coadapted.[26]

Actually observed changes with time in the frequency of hybridization under natural conditions have been claimed for species of the genus *Parus*.[7,27] However, Thielke[28] has clearly demonstrated just how flimsy this evidence is.

Other evidence[29,30] which is often considered to support the reinforcement alternative usually relates more to the parapatric meeting of the two diverged populations rather than the situation dealt with by Ayala et al. The parapatric situation has not been dealt with here in detail as it is somewhat distinct. However, it may be remarked that this evidence for the origin of specific positive assortative mating by reinforcing selection is also not compelling[6,31,32] and does not meet Moore's criticism[6] that alleles selected for their reinforcing action at the interface of the two populations will lose their selective value in allopatry. In fact it should be pointed out that they will generally be selected against, away from the interface, and not be neutral as Dobzhansky[33] has suggested. Selection against disadvantageous heterozygotes at a parapatric interface does not, in any case, seem likely to lead to speciation.[6,7,34]

A final line of reasoning is often invoked as indirect support for the reinforcement model.[35] It is pointed out that sexually dimorphic bird species such as the mallard (*Anas p. platyrhynchos*) and the pintail (*Anas a. acuta*) are usually from a continental fauna rich in species. Relatives from impoverished island faunas, in contrast, are frequently sexually monomorphic. In interpreting such observations it is often concluded that a causal correlation exists between the absence of close relatives and the sexual monomorphism. Unfortunately this explanation fails completely to explain why, for example, the five typical *Anas* species of the relatively rich South African duck fauna are all sexually monomorphic. The explanation for monomorphism and dimorphism in ducks is likely to have a basis unrelated to selection for reinforcement.[36]

CONSEQUENCES OF CLOSE COADAPTATION

Potential sexual partners of a particular species recognize each other by means of a specific-mate recognition system comprising a number of coadapted stages. In each of these a signal from one partner results in an appropriate response in the other, following the signal's reception and interpretation by the specifically coded central nervous system. Such a chain of coadapted stages[1,2] will, of course, be subject to intense stabilizing selection. Any mutation influencing a link in this chain to any significant extent will be selected against. Evidence exists which indicates that these processes in the two partners are under the control of genes at distinct loci.[37] (Recent reports[38,39] of "genetic coupling" of male signal-pattern-determining genes with female signal-pattern-recognition-determining genes provide little support for the view that the same genes determine both mechanisms.)

Each component of the SMRS is, of course, subject to a small, characteristic variance which permits selection of small variations to occur. Each such selective change must, of necessity, be followed by selection on the other partner to restore coadaptation. These properties enable an SMRS to be modified by selection to fit a new environment, for example. However, it is clear that such change must, inevitably, be slow. This constraint on speed of change imposes a critical handicap on reinforcement in its "race" with selection against the genetic or cytological basis [p. 371] of the hybrid disadvantage.[13] It is obvious that any circumstance which extends the time available for reinforcement to occur can only do so by reducing, reciprocally, the selective pressure for change. For example, it might be thought that a heterozygote with a relatively low selection coefficient such as $s = 0.1$ could provide a greater opportunity for reinforcement by virtue of the increased time required to eliminate the rarer chromosome arrangement (Fig. 2). This is not so because there is a corresponding reduction in the selective advantage of any reinforcing allele.

Mating partners are also coadapted with respect to preferred habitat and preferred breeding season. These species characters are thus also subject to stabilizing selection.

It is not at all evident that Ayala et al. or Bossert[13] took these awkward properties of coadapted characters into account in their models.

CONCLUSIONS

Dobzhansky[40] believed species to be "adaptive devices through which the living world has deployed itself to master a progressively greater range of environments and ways of living." This view imposes on its holder the obligation to accept that species are the *direct* product of selection, which, in turn, requires that the reinforcement model of speciation be accepted. A major difficulty is that no certain example of speciation by reinforcement can be pointed to, and the evidence supporting it is weak and indirect.

In sharp contrast, many examples are known which provide strong prima facie evidence for speciation in total allopatry. Species which have arisen in total allopatry from any closely related species cannot be the direct product of selection, and it follows that they can scarcely be regarded as "adaptive devices" of the sort pictured by Dobzhansky. These points are presented here to emphasize that the two principal models of speciation, speciation by reinforcement and speciation in total allopatry, would yield quite distinct sorts of species. Failure to accept the conceptual distinctness of these two kinds of species is a source of great ambivalence in, for example, Mayr's principal works. While, on the one hand, he rather constantly supported speciation in total allopatry, he has, on the other, supported Dobzhansky's concept of "isolating mechanisms." In doing this he has evidently not noticed that such characters, shaped by natural selection to serve the function of isolating a species reproductively from other species, can only arise according to the reinforcement model. It should be noted that Petersen's view of species which was quoted above is perfectly consistent with speciation in allopatry and does not invoke isolating mechanisms to prevent gene flow between gene pools.

Although the arguments presented above against the idea of speciation by reinforcement may be somewhat oversimplified, they at least take into account a number of important points which have previously been ignored or not noticed. As things stand, the objections to the model are very strong, and lead one to question why authors ever invoke the reinforcement model at all. Could the reason be that it is needed to justify preconceived ideas on the nature of species? If so, Crosby's words should be heeded:[15] "But integrity, the maintenance of an uncontaminated racial or subspecific genotype, has no intrinsic value. The consequences of loss of integrity may be good or bad according to circumstances, and it is the consequences we have to consider, not some hypothetical idealistic quality."

REFERENCES

1. Stitch, H. F. 1963. An experimental analysis of the courtship pattern of *Tipula oleracea* (Diptera). *Canadian Journal of Zoology* 41:99–109.
2. Mathews, R. W. 1975. Courtship in Parasitic Wasps. In *Evolutionary Strat-*

egies of Parasitic Insects and Mites, ed. P. W. Price, 66–86. New York: Plenum.

3. Bush, G. L. 1975. Modes of animal speciation. *Annual Review of Ecology and Systematics* 6:339–64.
4. Mayr, E. 1942. *Systematics and the Origin of Species*. New York: Columbia University Press.
5. Dobzhansky, T. 1951. *Genetics and the Origin of Species*. 3d ed. New York: Columbia University Press.
6. Moore, J. A. 1957. An Embryologist's View of the Species Concept. In *The Species Problem*, ed. E. Mayr, 325–38. Publication No. 50. Washington, D.C.: American Association for the Advancement of Science.
7. Mayr, E. 1963. *Animal Species and Evolution*. Cambridge, Mass.: Belknap Press of Harvard University Press.
8. MacArthur, R. H., and E. O. Wilson. 1967. *The Theory of Island Biogeography*. Princeton, N.J.: Princeton University Press.
9. Liebherr, J., and W. L. Roelofs. 1975. Laboratory hybridization and mating period studies using two pheromone strains of *Ostrinia nubilalis*. *Annals of the Entomological Society of America* 68:305–9.
10. Ayala, F. J., M. L. Tracey, D. Hedgecock, and R. C. Richmond. 1974. Genetic differentiation during the speciation process in *Drosophila*. *Evolution* 28:576–92.
11. Li, C. C. 1955. *Population Genetics*. Chicago: University of Chicago Press.
12. Whitten, M. J. 1971. Insect control by genetic manipulation of natural populations. *Science* 171:682–84.
13. Wilson, E. O. 1965. The Challenge from Related Species. Chap. 2 in *The Genetics of Colonizing Species*, ed. H. G. Baker and G. L. Stebbins, 7–27. New York: Academic Press.
14. Bazykin, A. D. 1969. Hypothetical mechanism of speciation. *Evolution* 23:685–87.
15. Crosby, J. L. 1970. The evolution of genetic discontinuity: Computer models of the selection of barriers to interbreeding between species. *Heredity* 25:253–97.
16. Key, K.H.L. 1974. Speciation in the Australian Morabine Grasshoppers: Taxonomy and Ecology. In *Genetic Mechanisms of Speciation in Insects*, ed. M.J.D. White, 43–56. Sydney: Australia and New Zealand Book Company.
17. Moran, C., and D. D. Shaw. 1977. Population cytogenetics of the genus *Caledia* (Orthoptera: Acridinae), III. Chromosomal polymorphism, racial parapatry and introgression. *Chromosoma (Berlin)* 63:181–204.
18. Lewis, H. 1961. Experimental sympatric populations of *Clarkia*. *American Naturalist* 95:155–68.
19. Mettler, L. E. 1957. Studies on experimental populations of *Drosophila arizonensis* and *Drosophila mojavensis*. *University of Texas Publication* 5721:157–81.
20. Koopman, K. F. 1950. Natural selection for reproductive isolation between *Drosophila pseudoobscura* and *Drosophila persimilis*. *Evolution* 4:135–48.

21. Wallace, B. 1950. An experiment on sexual isolation. *Drosophila Information Service* 24:94–96.

22. Knight, G. R., A. Robertson, and C. H. Waddington. 1956. Selection for sexual isolation within a species. *Evolution* 10:14–22.

23. Paterniani, E. 1969. Selection for reproductive isolation between two populations of maize, *Zea mays* L. *Evolution* 23:534–47.

24. Ehrman, L. 1971. Natural selection for the origin of reproductive isolation. *American Naturalist* 105:479–83.

25. Crossley, S. A. 1974. Changes in mating behaviour produced by selection for ethological isolation between ebony and vestigial mutants of *Drosophila melanogaster*. *Evolution* 28:631–47.

26. Petit, C., O. Kitagawa, and T. Takamura. 1976. Mating system between Japanese and French geographic strains of *Drosophila melanogaster*. *Japanese Journal of Genetics* 51 (2):99–108.

27. Vaurie, C. 1957. Systematic notes on palearctic birds, No. 26. Paridae: The *Parus caeruleus* complex. *American Museum Novitates* 1833:1–15.

28. Thielke, G. 1969. Geographic Variation in Bird Vocalizations: Their Relations to Current Problems in Biology and Psychology. Essays presented to W. H. Thorpe. Chap. 14 in *Bird Vocalizations*, ed. R. A. Hinde, 311–38. Cambridge: Cambridge University Press.

29. Littlejohn, M. J. 1969. The Systematic Significance of Isolating Mechanisms. In *Systematic Biology*, ed. C. G. Sibley, 459–82. Washington, D.C.: National Academy of Sciences.

30. Dobzhansky, T., and P. C. Koller. 1938. An experimental study of sexual isolation in *Drosophila*. *Biologisches Zentralblatt* 58:589–607.

31. Patterson, J. T., and W. S. Stone. 1952. *Evolution in the Genus* Drosophila. London: Macmillan.

32. Loftus-Hills, J. J. 1975. The evidence for reproductive character displacement between the toads *Bufo americanus* and *B. woodhousii fowleri*. *Evolution* 29:368–69.

33. Dobzhansky, T. 1940. Speciation as a stage in evolutionary divergence. *American Naturalist* 74:312–21.

34. Walker, T. J. 1974. Character displacement and acoustic insects. *American Zoologist* 14:1137–50.

35. Lack, D. 1974. *Evolution Illustrated by Waterfowl*. Oxford: Blackwell Scientific Publications.

36. Siegfried, W. R. 1974. Brood care, pair bonds and plumage in Southern African Anatini. *Wildfowl* 25:33–40.

37. Bastock, M. 1956. A gene mutation which changes a behavior pattern. *Evolution* 10:421–39.

38. Bentley, D. R., and R. R. Hoy. 1972. Genetic control of the neuronal network generating cricket (*Teleogryllus, Gryllus*) song patterns. *Animal Behaviour* 20:478–92.

39. Hoy, R. R., J. Hahn, and R. C. Paul. 1977. Hybrid cricket auditory behaviour: Evidence for genetic coupling in animal communication. *Science* 195:82–83.

40. Dobzhansky, T. 1976. Organismic and Molecular Aspects of Species Formation. In *Molecular Evolution*, ed. F. J. Ayala, 95–105. Sunderland, Mass.: Sinauer Associates.

NOTE

The substance of this paper formed the basis of the Invitation Lecture at the Sixth Congress of the South African Genetics Society at Pretoria in September 1977 and of a lecture to the Population Genetics Group of the British Genetical Society at Leeds in January 1978.

4　A Comment on "Mate Recognition Systems"

Paterson, H.E.H. 1980. A comment on "Mate Recognition Systems." *Evolution* 34:330–31.

Alan Templeton had seen a manuscript copy of my 1976 talk presented at the International Congress of Entomology in Washington (chap. 2) and mentioned it in a note in response to an earlier one by Ringo in which the question was asked: Why 300 species of Hawaiian Drosophila? *In his reference to my paper Alan mentioned the mate recognition system, but in doing so shaped it to fit his preexisting views, which were "isolationist" in style. The resulting conflation was profound, and I was obliged to do what I could to contain the problem. The exercise was worthwhile because it caused me to summarize the gist of my concept succinctly and to state my viewpoint in as strongly contrasting terms as I could. Of course, my firm statement was not a personal attack on Alan.*

In a recent response to a note by Ringo (1977) on "Why 300 species of Hawaiian *Drosophila*?" Templeton (1979) introduced a new aspect to the discussion. This was the idea of the mate recognition system, or, as I now prefer to call it, the specific-mate recognition system (SMRS), which I introduced in 1976. The SMRS is the essential component of a new concept of species, the recognition concept, which is conceptually quite distinct from the current paradigm, the isolation concept. According to Dobzhansky (1951, 1976), species under the latter concept are to be regarded as "adaptive devices" whose integrity is maintained through the possession of isolating mechanisms (Dobzhansky, 1935). Here I wish to comment on Templeton's use of the term "mate recognition system" because it differs fundamentally from my own. If it is not corrected at once, his usage may lead to confusion and misunderstanding when my concept is published in full.

According to the recognition concept, species are incidental consequences of adaptive evolution and cannot be regarded as "adaptive devices." The line of reasoning which leads to this conclusion runs as follows: In sexual species the achievement of syngamy is of fundamental significance. It is therefore not surprising that all sexual organisms possess adaptations which have evidently evolved to facilitate the achieving of fertilization (fertilization mechanisms). In each of these it is always possible to discern a subset, the specific-mate recognition system (SMRS). The SMRS comprises a coadapted signal-response chain of the sort illustrated by Stitch (1963) or, in more general terms, by Desmond Morris (1956). Although under strong stabilizing selection as I have explained previously (Paterson, 1976 [chap. 2], 1978 [chap. 3]), the SMRS is subject to modification under natural selection to improve its effectiveness should a small population become isolated in a new and distinct environment. Due to the stabilizing selection acting on the system, such adaptive change can only occur in small steps with coadaptation between male and female being reestablished at each step. The evidence for these statements comes from observations on the SMRSs of species in their preferred habitats (e.g., Morton, 1975). Speciation is said to have occurred when the SMRS of the members of the daughter population has been so extensively modified that it no longer functions effectively with members of the parental or any other population. It should be said that modification of the SMRS by natural selection may be indirect, being mediated by pleiotropy. Thus, speciation occurs as an incidental consequence of the adaptation of a small, isolated population to a new habitat. The SMRS, therefore, falls into the same category as other adaptive characters which are changed when a population is adjusted by natural selection to new conditions. This accounts for the common observation that modifications to the niche of members of a population generally accompany speciation (Mayr, 1963).

It is evident that a new SMRS, derived in this way, determines a new gene pool and, hence, a new species. According to the recognition concept, species are populations of individual organisms which share a common specific-mate recognition system (Paterson, 1978 [chap. 3]). Species are, thus, incidental effects of adaptive evolution.

If these arguments are fully appreciated it will be obvious how greatly my ideas have been misrepresented by Templeton when he wrote (1979:516), "The raison d'être of a mate recognition system is to [p. 331] prevent matings with other sympatric *Drosophila*." This should be stated as follows if it is to be in accordance with what I advocated at Washington in 1976: "The raison d'être of an SMRS is to ensure effective syngamy within a population occupying its preferred habitat." The difference revealed is fundamental, and, if thought through, might very well cast new light on "Why 300 species of Hawaiian *Drosophila*?"

I hope that this brief account will not be regarded as any form of complete statement of the recognition concept of species. A detailed account has been prepared for publication.

REFERENCES

Dobzhansky, T. 1935. A critique of the species concept in biology. *Philosophy of Science* 2:344–55.

———. 1951. *Genetics and the Origin of Species.* 3d ed. New York: Columbia University Press.

———. 1976. Organismic and Molecular Aspects of Species Formation. In *Molecular Evolution,* ed. F. J. Ayala, 95–105. Sunderland, Mass.: Sinauer Associates.

Mayr, E. 1963. *Animal Species and Evolution.* Cambridge, Mass.: Belknap Press of Harvard University Press.

Morris, D. 1956. The Function and Causation of Courtship Ceremonies. Fondation Singer-Polignac, Colloque Internationale sur l'Instinct (Paris, 1954): 261–87.

Morton, E. S. 1975. Ecological sources of selection on avian sounds. *American Naturalist* 109:17–34.

Paterson, H.E.H. 1976. The role of postmating isolation in evolution. Invited Lecture Fifteenth International Congress of Entomology, Washington, D.C., Symposium on the Application of Genetics to the Analyses of Species Differences. [This volume, chap. 2.]

———. 1978. More evidence against speciation by reinforcement. *South African Journal of Science* 74:369–71. [This volume, chap. 3.]

Ringo, J. M. 1977. Why 300 species of Hawaiian *Drosophila?* The sexual selection hypothesis. *Evolution* 31:694–96.

Stitch, H. F. 1963. An experimental analysis of the courtship pattern of *Tipula oleracea* (Diptera). *Canadian Journal of Zoology* 41:99–109.

Templeton, A. R. 1979. Once again, why 300 species of Hawaiian *Drosophila?* *Evolution* 33:513–17.

5

The Continuing Search for the Unknown and Unknowable: A Critique of Contemporary Ideas on Speciation

Paterson, H.E.H. 1981. The continuing search for the unknown and unknowable: A critique of contemporary ideas on speciation. *South African Journal of Science* 77:113–19.

I was sent Michael White's Modes of Speciation *for review. On reading it, I was frankly disappointed, finding it assertive and poorly argued. Writing a critical review was difficult because White had done much for me. It was on his recommendation that I was appointed to the Zoology Department at the University of Western Australia, which in turn led to my acquiring Australian nationality. Since 1976 there had been no contact between us, but I would certainly have preferred not to criticize the book. Another factor finally influenced my decision to write the review: I was somewhat disconcerted by the many favorable reviews of the work and the few critical comments in them.*

My extended review was used to provide a broader critique of practices in evolutionary writing, a number of concepts such as "instant speciation" by polyploidy and sympatric speciation. I also used it as a vehicle to introduce the classification of modes of speciation into two classes so that it became obvious that many models are no more than conceivable at best. We can be almost certain that species like the dodo and solitaire evolved in allopatry on the isles of Mauritius and Rodrigues, respectively. Accordingly, we can say in a rather loose way that allopatric speciation is a Class I model, and all the rest are no better than Class II models. I believe that this paper serves a more useful purpose than just providing a critique of White's book.

Discussions of the nature of speciation assume that species have an objective existence. If species cannot be objectively defined, and are merely

artificial constructs or subjective figments of the imagination of taxonomists, then speciation can hardly be said to be a real process. (M.J.D.
White, *Modes of Speciation*, p. 2)

In science, inadequate concepts and inconsistent logic often lie at
the root of disputation and disagreement. Since species biology has
become well known for its polemics, it is now essential to examine its
foundations and to look critically at our standards of argument in an
attempt at reducing disagreement. In this article I shall do this by
examining aspects of M.J.D. White's book *Modes of Speciation* (San
Francisco: W. H. Freeman, 1978, 455 pp).

Conceptually, this book belongs with those works of the last forty-
five years in which species are defined in terms of "reproductive
isolation." For this reason many of my comments apply as well to the
appropriate sections of other works such as Dobzhansky's *Genetics of
the Evolutionary Process*, Mayr's *Animal Species and Evolution*, or Grant's
Plant Speciation.

I owe much to Michael White, including that all-important encouragement received when young and much of my appreciation of
the role of chromosomes in evolution. My criticisms are, therefore,
to be seen as scientific, made in the hope that they will contribute to
the emergence of a theory of species which will refute Darwin's pes-
simistic view reflected in my title.

LOGICAL BACKGROUND

Mayr[1] was surely right when he wrote that "an understanding of the
nature of species . . . is an indispensable prerequisite for the understanding of the evolutionary process." Here he is clearly referring to
genetical, not taxonomic species. Concepts of species are found in two
very distinct fields of biological endeavor, taxonomy (the science of
classification) and population genetics. In both of these fields distinct
concepts of species are used, and their conflation is a major source
of confusion. When an author glides imperceptibly from talking of
species under one concept to talking of species in accordance with
another and distinct concept, a subtle kind of nonsense is generated,
which is exceedingly difficult to detect. None of the major authors
on evolution, including White, has avoided this flaw.

In discussing modes of speciation we are obliged to talk in terms
of a genetical concept, because speciation is a genetical event. I have
distinguished two quite distinct *genetical* concepts, the isolation concept
and the recognition concept.[2] Species, according to the former concept, are populations of interbreeding individuals "reproductively iso-

lated" from each other by "isolating mechanisms." Isolating mechanisms are thought of as safeguarding the species' "well-integrated, coadapted set of gene complexes." On the other hand, the recognition concept postulates that the limits of the gene pool of a species are determined, incidentally, by the members of the species mating positively assortatively as a result of their sharing a common specific-mate recognition system (SMRS).[2] Species are, thus, conceived of without reference to other species. The relational nature of the isolation concept is clear from Mayr's statement[3] (p. 319): "When no other closely related species occur, all courtship signals can 'afford' to be general, nonspecific and variable." Thus, these two genetical views of species are quite distinct, with distinct implications, and, yet, both have been very generally conflated by all the leading authors.

Logically, a book on speciation should open with a clear and comprehensive statement of the author's concept of species in genetical terms, for the reader needs to be able to judge for himself whether speciation has ever occurred in accordance with any particular model discussed. The preferred concept would lead to certain predictions which could then be used to evaluate the concept. The predictions would eliminate some modes of speciation from consideration.

Modes of speciation can be grouped into two classes: Class I, theoretical, with support from observed critical facts, and Class II, theoretical, without support from observed critical facts. Modes are assigned to Class I if the following question elicits an unconditionally positive answer: Do reliable, critical, observational data exist which compel acceptance of the view that speciation in accordance with this mode has actually occurred? All other modes are assigned to the obviously much less important Class II.

It is not sufficient to have each step of a model barely credible, for it is the compounded probabilities which are important. A series of barely credible steps will produce an improbable model as a whole. Class II models can be argued with impeccable logic and yet be impossible. This, for example, may be because the author assigns a role to an allele which is unrealistic. Therefore, when alleles with special properties are invoked as an essential part of a model, we should always ask: Are alleles such as these known to exist? if not, the model becomes so much less credible. We should do well to apply these simple rules to all modes of speciation which have been proposed.

In a book such as White's, and in a critical review such as this, it is inevitable that opposing views should be considered. This should be done objectively, for, as Popper[4] has pointed out, no service is done to science when a case under attack is presented in less than its strongest form.

These few general points are especially mentioned because they are frequently infringed by White (and others). Few authors appear sensitive to such flaws, and some are even unsympathetic. I found disconcerting the following statement made by White (p. 324) while discussing speciation: "It is emphatically not a field that should be encumbered by sterile semantic arguments as to the meaning and definition of terms, as has too often been the case in the past." I believe, on the contrary, that his book would have gained much in value had White provided a clear concept of what a species is, and then applied it consistently.

WHITE'S VIEWS ON THE NATURE OF SPECIES

White's first chapter does deal with "species and speciation," but his views are not stated very explicitly. On balance, however, it appears that they agree rather closely with those of Dobzhansky. He quotes Dobzhansky's definition of a species from 1970,[5] and one of Mayr's, though not that of 1970,[1] preferring for some reason the earlier one of 1940[6] and 1963.[3] Support for the biological species concept (essentially equivalent to the isolation concept in my terminology)[2] is implicit from the way he [p. 114] considers and discounts the attacks on the concept by Ehrlich and his colleagues and by Sokal and Crovello. However, the definitions he quotes refer to certain properties of species without making plain what the nature of species actually is. This he does, somewhat obliquely, on p. 14: "Speciation is one of the main ways by which living organisms adapt in order to exploit the diversity of environments available to them." Here again White is in essential agreement with Dobzhansky[7] (p. 104): "Species are not accidents but adaptive devices through which the living world has deployed itself to master a progressively greater range of environments and ways of living." These properties of a species bear little explicit relationship to Dobzhansky's[5] definition of a species: "Species are . . . systems of populations; the gene exchange between these systems is limited or prevented by a reproductive isolating mechanism or perhaps by a combination of several mechanisms," which says nothing about the role of species in nature.

At this point we must ask whether all the properties White has attributed to species are mutually compatible. In examining the definition of species quoted above, the problem at once ramifies when we look at the class "isolating mechanisms" and note that although they can all be described as "intrinsic" they are really a very disparate group of characters. Their influence on the exchange of genes between species, I believe, might very credibly be attributed to incidental consequences ("effects") of their true functions. Certainly, the so-called postmating isolating mechanisms could not have been selected to prevent gene exchange between species as Darwin showed at the beginning of chapter 8 of the *Origin of Species*. Furthermore, if White and Dobzhansky are correct in believing that species are adaptations selected to exploit available environments more effectively, then surely this can be achieved only by natural selection acting on characters related to this improved exploitation of nature? Among the premating isolating mechanisms the only possible candidates appear to be *habitat* and *seasonal isolating mechanisms*, for *ethological* and *mechanical isolating mechanisms* appear to have nothing whatever to do with this central property of species. Thus, White's concept of species and the definition he favors appear to be incompatible. This raises an awkward problem: How should we set about attempting to falsify his ideas in the Popperian manner? Which attributes of a species take precedence?

Perhaps the best that we can do in the circumstances is to combine the definition of Dobzhansky with the view that species are "adaptive devices evolved to exploit the diversity of environments available to them." If we do this, the following consequences follow:

1. Species are the direct products of natural selection, because they are adaptive devices which have evolved to exploit available environments.
2. Isolating mechanisms are thus true mechanisms (that is, adaptations) which have evolved to prevent gene exchange between distinct species.
3. Species comprise populations of sexual individuals.
4. Species are individual entities with genetic "integrity."
5. A species is at once a reproductive community and a gene pool.
6. Speciation, within these constraints, must occur in sympatry or, at least, in parapatry. In allopatry adaptation is all that is required to master a new environment.
7. It follows that natural selection in shaping isolating mechanisms acts "for the good of the species," as opposed to "the good of the individual."

In this essay I shall analyze White's arguments in relation to these stated or implied premises. It is not my object to assess the relative merits of the two genetical concepts of species.[2] In only a few places is it pointed out that a particular problem would not exist if an alternative species concept were adopted.

SPECIATION

Just as a clear view of species is needed in order to understand speciation, so clarity on how speciation occurs is mandatory if we are to grapple successfully with ecological concepts such as the niche and species diversity. In attempting to understand the development of a complex ecosystem, it is daunting to be obliged to take into account ten or more possible modes of speciation. In these circumstances species theory is effectively useless to population biologists. I strongly believe we should take only Class I models of speciation (see above) into account. This is important because our theory loses generality with every distinct mode of speciation admitted, and it seems unproductive to weaken it by considering seriously a model which may never have led to the formation of even a single species. I am, therefore, depressed when I find leading writers making rhetorical statements like "Speciation can occur by more than one way. Biologists who are too fond of simplicitly and homogeneity may be chagrined by the 'inventiveness' of the evolutionary process."[8]

Speciation to White is "the genetic changes whereby new species come into existence." He makes the important point which, indeed, needs emphasizing, that population geneticists "have not always seemed aware of the distinct genetic problems involved in speciation." He recognizes nine classes of speciation, but within each there may be distinct models, a formidable array. Which should we take seriously, which merely keep in mind as possibilities, and which reject?

Allopatric speciation
The overwhelming importance of allopatric speciation has been firmly established over the years, notably by Wagner,[9] K. Jordan,[10] D. S. Jordan,[11] and Mayr.[12] This is accepted by White (p. 107): "There can be absolutely no doubt about the reality of allopatric speciation . . . This is the one type of speciation whose actual existence is uncontroversial." However, he seems to believe that its status is somewhat *too* secure, for on p. 12 he states that acceptance of allopatric speciation has hardened into dogma. This appears to me to be rather uncharitable in view of the vast body of supporting evidence which appears to be incontrovertible. Allopatric speciation is, in fact, the only Class I model that we know of. Surely, then, it is reasonable to invoke it preferentially until another Class I model is demonstrated to exist? A stated aim of White's book is to loosen the dominance of the allopatric model, but, in fact, he provides us with only one Class I model.

That allopatric speciation is the only model of speciation which has sound credentials is a very awkward fact for all supporters of the isolation concept of species. Mayr[3] has written about isolating mechanisms: "They are *ad hoc* mechanisms. It is therefore somewhat difficult to comprehend how isolating mechanisms can evolve in isolated populations." This problem has existed since Wallace's day, though not all authors have been aware of it. In 1940[13] (p. 319) and in 1951[14] (p. 210) Dobzhansky attempted to resolve the dilemma as follows: "In such cases [of allopatric speciation] reproductive isolation might have arisen because these species have repeatedly invaded each other's territories; these attempts have led to the formation of isolating mechanisms." This is a slight modification of Wallace's[15] (p. 173) procrustean explanation of the same difficulty. Later, both Mayr[13] (p. 551) and Dobzhansky have attempted to play down the logical inconsistencies which exist between the isolation concept of species and the admitted occurrence of speciation in allopatry. Dobzhansky[5] (p. 376) wrote: "These two hypotheses [allopatric speciation and speciation by reinforcement] are not mutually exclusive. Needless disputes have arisen because they were mistakenly treated as alternatives." Despite his assertive words it should be obvious that the two models are indeed alternatives. During allopatric speciation, new species arise as an incidental consequence of adaptation to a new habitat. Species arising in this way cannot be regarded as adaptive devices in Dobzhansky's sense. In the case of speciation by reinforcement, natural selection has a direct role in forming the new species. Such species may be regarded as adaptive devices, depending on what reinforcement occurs. Suppressing or obscuring such a fundamental difference contributed [p. 115] significantly to the confusion which exists over the nature of species and speciation. There are, in many cases, doubts and difficulties over the model of speciation by reinforcement,[16,17] and it is certainly not a Class I model. Although White advocates the model periodically throughout the book, he is evidently not enthusiastic about it. On p. 152 he writes that "the known instances of 'reinforcement' [not 'speciation by reinforcement'] are still too few and too poorly understood for us to be certain whether it is a phenomenon of general importance." This is really quite surprising since the improvement of reproductive isolation through natural selection would appear, at first sight, to be the most obvious mode of speciation, and speciation in allopatry the least, if the isolation concept of species is valid (see the expectations derived from White's view of species above).

Clinal and area-effect speciation
Clinal and area-effect speciation are very much Class II models, no example of speciation according to them being known. Furthermore,

it is not very obvious why speciation should result from many of the situations described (e.g., the area effects in *Cepaea nemoralis* on the Wiltshire Downs and the rather mysterious case of *Maniola jurtina*).

Subspecies of a species are by definition[3] allopatric with respect to each other. It is rather surprising to find the following statement by White (p. 161): "A few evolutionists have gone beyond the claim that a complete barrier is a *sine qua non* only of successful speciation; they have actually extended this principle to include the formation of well-marked geographic subspecies between which there are no genetic isolating mechanisms."

Chromosomal models of speciation
Chapter 6 is devoted to the topic of chromosomal models of speciation and is clearly at the heart of the book. If White has insights for us on speciation, it is in this chapter that they are most likely to be found.

His particular contribution to speciation theory is the stasipatric model. It was advanced in 1967[18] and 1968[19] to account for the pattern of cytological variation which occurs in the *Vandiemenella viatica* complex of morabine grasshoppers of southern Australia.

White's Figure 16 summarizes the data: "The essential feature is a chromosomal rearrangement, originating somewhere within the area occupied by the ancestral species, which reduces fecundity when heterozygous. If such a rearrangement manages to establish itself (either by drift in a local deme or because it shows 'meiotic drive') it may spread geographically throughout a part of the area occupied by the species, because of homozygote superiority, and may act as an incipient isolating mechanism between the population homozygous for it and the original population" (p. 18). An alternative outline is provided on p. 177.

In considering this model, the problem of what White means by a species is very apparent. In this book, but not earlier, White recognizes no fewer than seven species in the *viatica* complex, though his close collaborator, Key,[20] recognizes only two when ostensibly applying Mayr's biological species concept (= isolation concept, essentially). It may be noted at once that neither drift nor meiotic drive involves natural selection.[21] This means that any species resulting from the postulated model of stasipatric speciation will not accord with Dobzhansky's or White's ideas on the nature of species which were discussed above. As with species formed in complete allopatry, species from stasipatric speciation are not adaptive devices. In this case they are the fortuitous consequences of cytological change leading to mechanical difficulties during meiosis. Does it matter that this lack of accord should exist between White's concept and his model of speciation? To me it must matter.

What if we were to accept Mayr's or Dobzhansky's definition of a species, but explicitly deny that species are adaptive devices (that is, that they are the direct products of natural selection)? Would stasipatric speciation, occurring as outlined by White, then constitute a plausible model of speciation?

There are still problems. For example, why are the $viatica_{17}$ and $viatica_{19}$ populations considered conspecific, together constituting *V. viatica*, and considered specifically distinct from P24(XY)? On Kangaroo Island P24(XY) is "reproductively isolated" from $viatica_{19}$ but not from the conspecific $viatica_{17}$; on the mainland $viatica_{17}$ and $viatica_{19}$ meet and hybridize along a parapatric zone. What concept of species is White applying here? *V. viatica*$_{17}$ appears to bear as comparable a relationship to P24(XY) as to $viatica_{19}$.

What is meant by the parapatric separation existing between many of the populations of this complex? Key[20] has shown intuitively, Moran and Shaw[22] have shown empirically, and it follows from the algebra of negative heterosis[17] that, at a parapatric zone of the sort existing in this case, the zone acts as a "semipermeable filter" which holds up the rearrangements at the interface while allowing shared chromosomes to pass through. In other words, the boundaries between the populations are demarcated by the distribution of the rearrangements only, and gene exchange involving loci of all other linkage groups occurs more or less freely. Furthermore, if it is borne in mind that the gene pools of two parapatric populations can conceivably be effectively identical, one is forced to enquire about what exactly is being guarded by this particular set of so-called isolating mechanisms. It can be seen that even with this less stringent view of species the stasipatric model is unsatisfactory, as it yields a kind of species which is rather meaningless in evolutionary terms. In fact, all species which are "isolated" by postmating "mechanisms" alone will be found to be unsatisfactory because, even when the isolation involves complete sterility or inviability of the hybrids, coexistence between such species cannot occur. This is because, in the absence of positive assortative mating, the smaller of the two populations will inevitably become extinct.[23,17] It is of interest that White does not invoke reinforcement at the parapatric zones in his model. It is, therefore, not clear how the "incipient isolating mechanisms" are extended.

To sum up: Speciation according to the stasipatric model may never have occurred. The model, in any case, would not yield species which accord with White's or Dobzhansky's concept of species. Even with a less stringent concept of species than that accepted by these authors, the model is not likely to be effective. At best it would have to be classified as a Class II model. The spreading of a chromosome rearrangement which is disadvantageous in the heterozygous state is not in itself speciation; otherwise species could easily be created in the laboratory. It is hazardous to infer from the present-day distribution of the *viatica* group, in an area much disturbed by agriculture, that the rearrangements which characterize the different populations arose as the model predicts. They could well have arisen in small relic populations cut off by Pleistocene changes in climate. Fixation by stochastic events would then be favored. The small, isolated populations would expand and spread following amelioration of the climate, eventually meeting and forming parapatric zones. Lande[24] has recently studied the matter theoretically.

Harlan Lewis[25] and his colleagues have suggested a somewhat similar model which has been called saltational speciation and which has been related to "catastrophic selection"[26] (extremes of climate). The angiosperm species *Clarkia lingulata* is supposed to have arisen from the widespread *C. biloba* in a tiny isolate.[25] In this case the several chromosomal rearrangements induce a great reduction in fecundity. Sympatry has been experimentally contrived,[27] and, as expected from population genetic algebra, the smaller population was generally soon eliminated, crossing being at random, since both populations possess a common specific-mate recognition system. The same problem over concepts of species that was found to occur with stasipatric speciation applied in this case as well: the resulting "species" are unusual since sympatry between daughter and parent populations is unstable, and the species cannot be regarded as adaptive devices. In fact, Lewis (p. 220) believes *lingulata* to be disadvantageous. In passing, it may be noted that applying the [p. 116] recognition concept of species to these *Clarkia* populations leads to the conclusion that *lingulata* and *biloba* are conspecific since they share a common specific-mate recognition system,[17] a conclusion which I have little doubt ecologists and taxonomists would endorse.

There is no doubt that closely related species often differ from each other by fixed alternative chromosome arrangements. There certainly seems to be a correlation between speciation and the fixation event, but there is no reason to believe that the fixation of the rearrangement has *caused* the speciation to occur; it is more probable that a population bottleneck occurred which simultaneously favored speciation and the fixation of the rearrangement.

White certainly seems to be over-enthusiastic in his advocation of an important role for chromosome rearrangements in evolution. On

p. 215 he concludes that the $n = 13$ population of *Nucella lapillus* which occurs on the exposed rocks of the Britanny coast (and elsewhere) is adapted to its severe habitat to some extent by advantageous properties of the five chromosome fusions, which distinguish this population from the $n = 18$ populations which occur in sheltered bays. Again there is a correlation, but attributing an adaptive role to the fusions is not justified without external evidence. On p. 225 he expresses the opinion that it now seems abundantly clear that chromsomal rearrangements possess an overwhelming importance in speciation; however, the evidence for this remains doubtful.

Sympatric models of speciation
In 1969 Bush[28] wrote of *Rhagoletis* that, "although some form of allopatric speciation can be invoked to explain the origin of all the currently sympatric sibling species in this genus, other modes of speciation appear to be more plausible when the known biological characteristics of this group of flies are considered."

For a decade Bush has been an advocate of sympatric speciation. For this he deserves credit because he has been attempting to overthrow a major paradigm. He has succeeded in winning the support of a number of leading evolutionists, including Michael White, with whom he has long been associated. In fact, it is certainly time to look seriously at what Bush is advocating and to decide whether he is at all justified in asserting that sympatric speciation is more plausible than the allopatric mode in accounting for certain speciation events, notably those in the tephritid fly genus *Rhagoletis* to which his own research has been devoted.

The first point to be made is that Bush himself has accepted that the model of sympatric speciation which he advocates is not a Class I model; it is not *certain* that speciation has actually ever occurred in accordance with its strictures. The best he can do is to argue that the sympatric speciation model is more plausible than the allopatric model. Clearly, the judging of plausibility is subjective, and the argument is scarcely compelling. Let us examine his case.

In discussions such as the present one we must be quite clear on the concept of species which is being used. Bush[29] (p. 339) has written, "Most evolutionists . . . generally accept as a working definition the *biological species concept*: species represent groups of interbreeding natural populations reproductively isolated from other such groups." Thus, Bush and White agree on the nature of species, both subscribing to the isolation concept. By way of introduction to Bush's work, White writes, "The sympatric origin of host races in *Rhagoletis*—and hence of species, because there can be no reasonable doubt that these host races are potentially incipient species—seems to be as firmly established as any case of allopatric speciation has been." It is essential to pause and analyze this peculiarly complex sentence word by word, because in it White aims at establishing a connection between Bush's work on host-range extension in *Rhagoletis pomonella* and sympatric speciation. Glide past it uncritically and the essential fallacy in White's (and Bush's) argument will have been missed.

First, we need to know what is meant by a host race. Bush pointed out that the term has been used in different ways by different authors, but then he writes:[30] "I will limit it to an infraspecific category generally applied to populations of a parasitic species which exhibit distinct genetically-based preferences for certain host plants." We should then ask whether the *Rhagoletis pomonella* individuals which now utilize apples as breeding hosts (instead of the normal hawthorn host) meet the definition of "host race." An alternative hypothesis is that *R. pomonella* has simply extended its already wide range of hosts[31] to include cultivated apples. To establish that the flies utilizing apples constitute a host race in Bush's sense, it is necessary to establish that the choice of the apple host is genetically determined. I am not aware that evidence establishing this has been provided, and so at present we cannot choose between the two possibilities. Even so, White's sentence is not acceptable because I fail to understand how sympatric host race formation can be made equivalent to speciation. According to Bush's definition, host races are infraspecific categories. This means, for example, that the apple flies still share a common gene pool with the hawthorn flies. In turn, this means that they are not intrinsically reproductively isolated from the hawthorn flies. Evidence on gene flow is not provided. We do not know whether late-emerging cherry flies are ever attracted to early apples, or whether late-emerging flies are attracted to early hawthorn fruits in Door County, Wisconsin (ref. 31, Fig. 2).

The final clause of the sentence is also doubtful. White's insecurely based assertions appear to stand in sharp contrast to the basis for accepting allopatric speciation.[3] As already noted, White himself accepted that there can be absolutely no doubt about the reality of allopatric speciation. Thus, in my view, White fails to establish any necessary connection between the extension of host range of *R. pomonella* and speciation.

The model of sympatric speciation which Bush[30] has offered, and which is rather inadequately outlined by White, is also open to criticism. It involves two loci; one is the host-determining locus (H), and the other determines larval survival on the fruit of the host (S). The author makes much of the point that only two loci are involved, implying that speciation occurs without a "genetic revolution." Attention to detail reveals that Bush is confused. The model is sympatric speciation, yet it is by no means obvious how its products differ from Bush's "host races," as defined above. Host selection and survival on the host are supposed to be genetically determined in both, but the host race is explicitly called an "infraspecific category"; which then determines the specific attributes of the product of the model? I confess to being baffled and wonder why others are not. The sentence from White's book which was analyzed above seems to reflect the state of Bush's argument very accurately!

There are more problems with the model. First, the evidence adduced for the simple basis of host choice and survival on the host is far from compelling. However, let us accept it as credible for now, and proceed to look at the rest of the argument. H_1 is the allele which determined hawthorn as the host, and H_2 the allele determining preference for the new host. Similarly, S_1 determined survival on hawthorn, while S_2 determined survival on the new host. It seems that preference for, and survival on, hawthorn are dominant characters. If we assume mutation rates of 10^{-5} at both loci, we have a compounded probability of 10^{-10}, which at once begins to provide perspective. Of course, the new mutants are unlikely to occur together in one small deme. It is not at all clear why the heterozygotes should survive or the new alleles spread because $H_1H_2;S_1S_1$, $H_1H_2;S_1S_2$, and $H_1H_1;S_1S_2$ should have the same adaptive values as the parental homozygotes. The new alleles under these circumstances are very likely to be lost purely by chance,[32] and there is no obvious reason for them to spread. If spread by drift is to be invoked, a small population must be specified. This makes it very improbable that both mutation events will occur in it simultaneously. To spread the alleles in a large population will require invoking either heterozygote advantage (which, in turn, has important disadvantages), meiotic drive (a rare phenomenon [p. 117] unlikely to occur for two distinct chromosomes in one population), or through close linkage to an advantageous new allele at another locus which is in the process of spreading toward fixation (again an improbable event, unlikely to oblige at both loci).

In a population which has two alleles segregating independently at each of two loci, nine genotypes are possible. According to the assumptions of Bush's model, four of these genotypes are inviable on either host; four are viable on hawthorn; and only one ($H_2H_2;S_2S_2$) will find and survive on the new host. Even so, at least a male and a female of this genotype must occur simultaneously and then choose the same tree of the new host out of all those available. They must meet and mate successfully, the female must oviposit, and the larvae must survive to yield a second generation. Compounding all these probabilities demonstrates that the generation of a new host race in accord with the model is unlikely in the extreme.

But, it may be protested, it has been observed to happen in *Rhagoletis pomonella*. It must be pointed out that we do not know that the extension of host range from hawthorn to apple or cultivated cherry occurred in this way, and we do not know that gene exchange does not occur between the flies on the different hosts where they occur sympatrically as in Door County, Wisconsin. It is possible that no genetic change was involved. Bush[31] has published a range of natural hosts of *R. pomonella*. These comprise fifteen species of *Crataegus*, three of *Pyrus*, three of *Prunus*, and one of *Cotoneaster*. No data appear to be available about gene flow between these populations, or the ability of larvae from one host to survive on another species. As Bush points out, new cultivated hosts are more easily colonized than are new hosts in nature, because they have often lost their protective chemicals during the process of selection for desirable characteristics. Bush makes much of the differences in peaks of abundance of flies from different hosts, but we do not know how these were determined. The diagrams illustrating this phenomenon certainly show considerable overlap. Again, we do not have data on gene exchange between cohorts, and it would be difficult to obtain it in any direct manner.

Sufficient has now been written to show that it is very improbable that sympatric speciation can occur according to Bush's model. This model was selected from the innumerable candidates because it has had more influence than the others in convincing many evolutionists, including White, that sympatric speciation should be considered seriously. White has discussed many other cases which he believes can be explained only in terms of sympatric speciation. However, I fail to see how any case of speciation can ever be attributed to sympatric speciation in the confident way we can attribute cases to allopatric speciation. For example, White makes out a case for the weevils of the genus *Microcryptorhynchus* on the remote and minute oceanic island of Rapa. What appears to be a convincing case to White is less convincing to others such as Carson.[21] If examples of speciation in sympatry cannot be recognized without ambiguity then there is no possibility of ever testing a model of sympatric speciation.

White ends his chapter: "However, in spite of a rather woeful lack of evidence as to the genetic processes involved, it seems impossible today to deny the reality of sympatric speciation, at least in many groups of insects." I am afraid I cannot accept this assertion. I believe that, rationally, we must keep looking for compelling evidence that sympatric speciation has actually ever occurred, for only then will one of the many models of sympatric speciation move from Class II to Class I. At present, I for one am yet to be convinced that sympatric speciation has ever occurred.

SPECIATION BY POLYPLOIDY

Polyploidy is of special interest in the study of speciation. It provides a sharp test for different concepts of species by posing awkward questions to their adherents. Authors such as Mayr, Dobzhansky, and White, who support the isolation concept of species, are inclined to write as follows: "The production of a polyploid constitutes instantaneous speciation; it produces an incompatibility between the parental and the daughter species in a single step."[3] The argument continues: "As polyploids and parental population are genetically isolated the two may evolve along divergent paths. The polyploid usually comes to occupy a different habitat from the diploid and may acquire characters that are so different that it must be classified as a separate species."[33]

White devotes a chapter to speciation by polyploidy, but in his first chapter he draws attention to the fact that Mayr,[34] while accepting twelve models of speciation as conceivable, concluded that, besides allopatric speciation, only speciation by polyploidy was of real importance. In this section I shall review briefly the reasons for such a conclusion by Mayr and then see if they are justifiable.

It seems to me that polyploidy is assigned an important role in speciation for two main reasons. First, in nature, particularly among plants, many species differ from near relatives in their level of ploidy. Secondly, the occurrence of such pairs of sister species can at once be explained in terms of polyploidization, which ostensibly yields a new species instantly in accord with the isolation concept of species.

In considering two sister species we usually have no way of knowing the particular course taken by speciation. If, for example, the two species differ by fixed alternative inversions or translocations, we cannot be certain whether these cytological differences have had a direct causative role in speciation, or whether they correlate with the speciation event quite incidentally, though the latter is more probable. Similarly, with differences in levels of ploidy, was the daughter species formed instantly by polyploidization, or by a small tetraploid subpopulation speciating allopatrically after dispersing into a distinct, isolated habitat? It is curious that this latter possibility is never considered. Perhaps it is due to the first explanation having been considered so overwhelmingly self-evident. If this is the reason, it should be challenged.

Careful review[35] has revealed that in plants few autopolyploids are to be found in nature. Why should this be so? A sexual autotetraploid which arises within a population of diploid plants can persist only under certain conditions. It must be capable of self-pollination or some form of asexual reproduction. An obligatory outbreeder without access to any form of asexual reproduction will be eliminated. A tetraploid which can exploit both self- and cross-fertilization will probably be eliminated because of the preponderance of pollen from diploid plants. Any autotetraploid which relies entirely on asexual reproduction falls outside the definitions of species of authors such as Dobzhansky and Mayr, which refer to sexual organisms. There are other reasons why autotetraploids do not thrive and why they are rare in nature. They are very much slower than diploids in responding to selection.[33] They are also subject to a number of disadvantages. Meiosis is irregular in a newly arisen tetraploid and stabilizes only after some time, possibly due to natural selection. They are often susceptible to frost because of the commonly reduced osmotic pressure of their cell sap.[33] An autotetraploid is not likely to be perceptibly different from the parental diploids in most adaptive characters since its genes are drawn from the diploid's gene pool. When White (p. 261) calls polyploids "biotypes," and when Dawson[33] (see above) says, "the polyploid usually comes to occupy a different habitat from the diploid," they are clearly referring to allopolyploids. Because of their similar genotypes, autotetraploids are generally unable to escape from contact with their parental diploid. For such reasons it is not surprising that autotetraploids are uncommon. In fact, so few natural autopolyploids are known, little can be said about their properties. Most detail comes from those artificially produced.

Stebbins[35] and others have made it quite clear that among plants in nature most polyploids are allopolyploids—tetraploids of hybrids of two subspecies or species. Such allotetraploids differ in genotype from [p. 118] individuals of either parental population. Granted reasonably efficient dispersal mechanisms, they may well come to occupy a different habitat from either parental population. In allopatry the allotetraploid may well survive and slowly adapt to the conditions of the new habitat, achieving in due course full specific status. However, this will not then be a case of instant speciation by polyploidization, but speciation in allopatry. Allotetraploids which remain in contact with one or other parental diploid population will be subject to the same problems as is the autotetraploid.

That postmating "reproductive isolation" is not a satisfactory basis for defining species can be seen from the example of autotetraploids. It is little wonder that plant taxonomists have often refused to pay attention to polyploids which are indistinguishable in phenotype from the diploids.[36] "Such a form is often referred to as a 'polyploid race,' as by Seiler in *Solenobia*. Yet such a 'race' is reproductively isolated from the parental species and is, biologically speaking, a good species. This is another illustration of the frequent conflict between a morphological and a biological species concept."[3] It is perhaps appropriate to point out that Mayr is talking about the "isolation concept" when he writes "biological concept," and that if the "recognition concept," which is just as biological, is applied in place of it, the conflict does not exist!

Cronquist[36] has recently drawn attention to the problems which arise by applying the "isolation concept" to plants among which sterility is often an *intraspecific* phenomenon. He writes: "Self-sterile plants are usually also sterile with a certain fraction of the population to which they belong." Similar phenomena occur among animals. In the Llanos A stock of *Drosophila paulistorum* of the so-called Orinocan semispecies, Dobzhansky and Pavlovsky[37] found that it was at first (1958) fully interfertile with other Orinocan strains, but when retested in 1963 it was found to be intersterile with all other Orinocan stocks with which it had formerly been interfertile. Similar changes in interfertility have been recorded in mosquitoes within the species *Culex molestus*.[38] Members of the *Drosophila paulistorum* complex carry symbiotic mycoplasms in their cytoplasm,[39] which are coadapted with the host's genotype, while in *C. molestus* there is a comparable rickettsial symbiont.[40] Such systems appear to be very prone to developing intercolony sterility under laboratory conditions.[38]

Thus sterility and, for that matter, other "postmating isolating mechanisms" are not satisfactory criteria for delineating species. This well-known, but much ignored, point can be further illustrated by considering polyploids. Dobzhansky and White have regarded species as adaptive devices. This view cannot be sustained with newly arisen polyploids, for, as White himself accepts (p. 128), "postmating isolating mechanisms" are acquired as "effects," and cannot be regarded as adaptive. With respect to polyploids, conventional views on the role of "isolating mechanisms" and the properties of species seem to be forgotten. Mayr[3] (p. 109) expressed the common viewpoint: "It is the function of the isolating mechanisms to prevent such a breakdown and to protect the integrity of the genetic system of species." In writing on polyploidy and speciation, many authors accept that most natural polyploids are allopolyploids which arose by "instantaneous speciation," but fail to notice that they are now accepting that an "isolating mechanism" is protecting a hybrid genotype, the very thing "isolating mechanisms" are supposed to prevent! The inadequacies of "reproductive isolation" as a criterion for recognizing species will be returned to below.

The following conclusions about polyploids seem to be justified. Autopolyploidy should be regarded as an intraspecific phenomenon since autopolyploids share a common specific-mate recognition system with the parental diploids and so crossing would occur at random with respect to them. Newly arisen allotetraploids should be treated in the same way as other hybrids. Good species may arise from allo-tetraploids by allopatric speciation (acquiring a unique specific-mate recognition system by adapting to a new habitat to which they have been restricted by a chance event).

ASEXUAL SPECIATION

It was shown above that White's views on species are close to those of Dobzhansky in most respects. However, these authors differ over the treatment of asexual organisms. Dobzhansky[5] (pp. 357–58) has written that "to talk of reproductive isolation is meaningless in asexual, parthenogenetic or obligatory self-pollinating forms, and yet systematists name species everywhere . . . Species is not only a category of classification but also a form of supra-individual biological integration. In the former sense, any taxon may be called a species if it is convenient to use this designation. Species then become as arbitrary as subgenera, genera and other categories." It is evident that Dobzhansky is making a clear distinction between taxonomic and genetical concepts of species and applying the former to asexual, and the latter to sexual, organisms. This is logical and pragmatic provided that it is remembered that genetical concepts are part of the field of population genetics. Decisions taken there can, of course, then be related to the field of taxonomy.

Mayr,[3] although somewhat inconsistent, clearly supported Dobzhansky's use of a purely taxonomic concept for asexual organisms when he wrote that "the word 'species' signifies not only the biological unit of a reproductively isolated population, but also the classifying unit of a kind of organism. It is perhaps this consideration that has induced most practicing taxonomists to be frankly dualistic: they define the term species biologically in sexual organisms and morphologically in asexual ones."

White justifies discussing asexual speciation in chapter 9 because "the majority of asexual populations seem to have been derived rather recently from sexually reproducing forms and that consequently their genetic systems bear the imprint of their origin, so that in most cases remnants of an organization into biological species clearly persist."

He continues, "It is clear that whenever a population that reproduces by vegetative or parthenogenetic processes arises from a sexually reproducing one, an evolutionary event has occurred that involves reproductive isolation, the severance of genetic continuity. It seems better to extend the meaning of the term 'speciation' to cover such events than to invent another term."

White's lack of concern for the meaning of terms undermines this whole book. Of course, it is not just the definition of the term "speciation" which is affected, but the concept of species as well. However, the whole argument fails to stand up to examination. It should be noticed that White speaks of an evolutionary event which gives rise to "a population that reproduces by vegetative or parthenogenetic processes." This is obviously not the case. It is asexual *individuals* which are formed, each part of an independent clone in which the only transmission of genes that occurs is from unisexual parent to unisexual offspring. Each line or clone of an asexual taxon is as isolated from any other as it is from the sexual species from which it arose. I fail to understand how White can accommodate his views on asexual species with his statement (p. 3), "It cannot be emphasized too strongly that every species is at the same time a *reproductive community*, a *gene pool*, and a *genetic system*" (White's italics).

Regardless of which genetical concept of species is adhered to, tolerance of "asexual species" involves an unacceptable logical distortion. It is just such ad hoc tampering with definitions and concepts which is undermining modern theories of population biology.

Many obligatory asexual organisms in nature, like most polyploids, appear to have a hybrid origin. The adoption of asexuality, or polyploidy, thus, may often be an "escape" from the disadvantages of hybridity. This was pointed out long ago by Darlington.[41] This may be the case with [p. 119] *Warramaba virgo* so extensively studied by White. In such cases, the permanent heterozygosity that occurs can be identified as an effect rather than adaptation.[42]

DISCUSSION

Evolutionary biology is a complex subject which demands the integration of information from many disciplines, each normally independent and possessing its own structure, concepts, methods, and burgeoning literature. A modern synthesis requires an overall familiarity with this vast field. Furthermore, "data of experience" from these sources are to be used to test alternative evolutionary paradigms and hypotheses. To do this effectively calls for the observance of strict standards of logical analysis and constant attention to the dangers of bias and prior commitment. Another danger that needs identifying involves the shortcomings of language as a vehicle for complex ideas: the speciousness of an argument can be kept unapparent by a particular choice of words.

An unwarranted complacency exists over the quality of argument which is currently accepted in evolutionary discussion. I have attempted to demonstrate this by examining a recent book by a noted author, for I believe that it is vital that we become more critical if we are to resolve present-day disagreements.

It is apparent that White adhered to the isolation concept of species. To paraphrase Simpson,[43] it is, thus, the known consequences of the genetical situation stated in this concept which determine the limits within which White's modes of speciation must lie. A model which survives these tests cannot yet be considered established; it must also be shown to be realistic. This is a difficult thing to do, since we have only a limited capacity to take into account all the factors likely to be of importance under natural conditions. This is why Class I models are so important: the only reliable test of realism is the fact that evidence exists that a species has in fact arisen in accord with the constraints of the model.

At the end of White's book I remained unconvinced that he has established the existence of any Class I model other than allopatric speciation. It is true he made many assertions to the contrary, but these were not convincingly supported by the facts. For example, he ended his chapter on sympatric speciation with the remark that it seemed impossible today to deny the reality of sympatric speciation, a view which can be seen in perspective by noting that even Bush accepted that it was not possible to distinguish with certainty between a supposed case of sympatric speciation and an allopatric speciation alternative. In the case of other modes, there are either flaws, since the observed facts are not in accord with White's concept of species, or there is an absence of evidence that speciation has ever occurred in accord with the constraints of the mode, so that, at best, they must be assigned to Class II.

Other major texts share the logical and semantic flaws of this book. Perhaps enough has been done to show that many ideas in population biology are insecurely based and call for the systematic attention of specialists in experimental design. Which ideas are firmly based, and which held because they agree with deeply seated preconceived ideas? Different modes of speciation follow from different concepts of species. At present a new species concept is being sought. If it emerges, a new review of speciation will be needed.

I thank my colleagues and students for hours of testing discussion.

In particular, I am indebted to Marc Centner, Robin Crewe, Christopher Green, Stuart Halse, David Lambert, Judith Masters, Neville Passmore, Hugh Robertson, and Elisabeth Vrba.

REFERENCES

1. Mayr, E. 1970. *Populations, Species and Evolution.* Cambridge, Mass.: Belknap Press of Harvard University Press.
2. Paterson, H.E.H. 1980. A comment on "Mate Recognition Systems." *Evolution* 34:330–31. [This volume, chap. 4.]
3. Mayr, E. 1963. *Animal Species and Evolution.* Cambridge, Mass.: Belknap Press of Harvard University Press.
4. Griffiths, G.C.D. 1973. Some fundamental problems in biological classification. *Systematic Zoology* 22:338–43.
5. Dobzhansky, T. 1970. *Genetics of the Evolutionary Process.* New York: Columbia University Press.
6. Mayr, E. 1940. Speciation phenomena in birds. *American Naturalist* 74:249–78.
7. Dobzhansky, T. 1976. Organismic and Molecular Aspects of Species Formation. In *Molecular Evolution*, ed. F. J. Ayala, 95–105. Sunderland, Mass.: Sinauer Associates.
8. Dobzhansky, T. 1972. Species of *Drosophila. Science* 117:664–69.
9. Wagner, M. 1889. *Die Entstehung der Arten durch raumliche Sonderung.* Basel: Benno Schwalbe.
10. Jordan, K. 1905. Der Gegensatz zwischen geographischer und nichtgeographischer Variation. *Zeitschrift für Wissenschaftliche Zoologie* 83:151–210.
11. Jordan, D. S. 1905. The origin of species through isolation. *Science* 22:545–62.
12. Mayr, E. 1942. *Systematics and the Origin of Species.* New York: Columbia University Press.
13. Dobzhansky, T. 1940. Speciation as a stage in evolutionary divergence. *American Naturalist* 74:312–21.
14. Dobzhansky, T. 1951. *Genetics and the Origin of Species.* 3d ed. New York: Columbia University Press.
15. Wallace, A. R. 1889. *Darwinism.* London: Macmillan.
16. Moore, J. A. 1957. An Embryologist's View of the Species Concept. In *The Species Problem*, ed. E. Mayr, 325–38. Publication No. 50. Washington, D.C.: American Association for the Advancement of Science.
17. Paterson, H.E.H. 1978. More evidence against speciation by reinforcement. *South African Journal of Science* 74:369–71. [This volume, chap. 3.]
18. White, M.J.D., R. E. Blackith, R. M. Blackith, and J. Cheyney. 1967. Cytogenetics of the *viatica* group of morabine grasshoppers, I. The "coastal" species. *Australian Journal of Zoology* 15:263–302.
19. White, M.J.D. 1968. Models of speciation. *Science* 159:1065–70.
20. Key, K.H.L. 1974. Speciation in the Australian Morabine Grasshoppers: Taxonomy and Ecology. In *Genetic Mechanisms of Speciation in Insects*, ed.

M.J.D. White, 43–56. Sydney: Australia and New Zealand Book Company.

21. Carson, H. L. 1974. Chromosomes and species formation. *Evolution* 32:925–27.

22. Moran, C., and D. D. Shaw. 1977. Population cytogenetics of the genus *Caledia* (Orthoptera: Acridinae), III. Chromosomal polymorphism, racial parapatry and introgression. *Chromosoma (Berlin)* 63:181–204.

23. Li, C. C. 1955. *Population Genetics.* Chicago: University of Chicago Press.

24. Lande, R. 1979. Effective deme sizes during long-term evolution estimated from rates of chromosomal rearrangement. *Evolution* 33:234–51.

25. Lewis, H. 1966. Speciation in flowering plants. *Science* 152:167–72.

26. Lewis, H. 1962. Catastrophic selection as a factor in speciation. *Evolution* 16:257–61.

27. Lewis, H. 1961. Experimental sympatric populations of *Clarkia*. *American Naturalist* 95:155–68.

28. Bush, G. L. 1969. Sympatric host race formation and speciation in frugivorous flies of the genus *Rhagoletis* (Diptera: Tephritidae). *Evolution* 23:237–51.

29. Bush, G. L. 1975. Modes of animal speciation. *Annual Review of Ecology and Systematics* 6:339–64.

30. Bush, G. L. 1974. The Mechanism of Sympatric Host Race Formation in the True Fruit Flies (Tephritidae). In *Genetic Mechanisms of Speciation in Insects*, ed. M.J.D. White, 3–23. Sydney: Australia and New Zealand Book Company.

31. Bush, G. L. 1975. Sympatric Speciation in Phytophagous Parasitic Insects. In *Evolutionary Strategies of Parasitic Insects*, ed. P. W. Price, 187–206. London: Plenum.

32. Fisher, R. A. 1958. *The Genetical Theory of Natural Selection.* 2d ed. New York: Dover Publications.

33. Dawson, G.W.P. 1962. *An Introduction to the Cytogenetics of Polyploids.* Oxford: Blackwell.

34. Mayr, E. 1957. Die denkmöglichen Formen der Artentstehung. *Revue Suisse de Zoologie* 64:219–35.

35. Stebbins, G. L. 1974. Types of polyploids: Their classification and significance. *Advances in Genetics* 1:403–29.

36. Cronquist, A. 1978. Once Again, What Is a Species? In *Biosystematics in Agriculture*, ed. J. A. Romberger, 3–20. New York: Wiley.

37. Dobzhansky, T., and O. Pavlovsky. 1966. Spontaneous origin of an incipient species in the *Drosophila paulistorum* complex. *Genetics* 55:727–33.

38. Irving-Bell, R. J. 1977. The biology of the ovary in the *Culex "pipiens"* complex (Diptera, Culicidae). Ph.D. diss., University of Western Australia.

39. Ehrman, L., and R. P. Kernaghan. 1971. Microorganismal basis of infectious hybrid male sterility in *Drosophila paulistorum*. *Journal of Heredity* 62:67–71.

40. Hertig, M. 1936. The rickettsia, *Wolbachia pipientis* (gen. and sp.n.) and

associated inclusions of the mosquito, *Culex pipiens. Parasitology* 28:453–90.

41. Darlington, C. D. 1958. *Evolution of Genetic Systems.* 2d ed. Edinburgh: Oliver and Boyd.
42. Williams, G. C. 1966. *Adaptation and Natural Selection.* Princeton, N.J.: Princeton University Press.
43. Simpson, G. G. 1961. *Principles of Animal Taxonomy.* New York: Columbia University Press.

NOTES

The author is professor and head of the Department of Zoology, University of the Witwatersrand, Johannesburg 2001, South Africa. This article is Publication 1 of the Program on Animal Communication, financed by the Council Research Committee of the University of the Witwatersrand.

Note added in proof [in 1980]: A paper with some aims comparable with those of this article has recently been noticed. Futuyma, D. J., and G. C. Mayer. 1980. Non-allopatric speciation in animals. *Systematic Zoology* 29: 254–71.

6   Epitaph to a Scientific Gadfly

Paterson, H.E.H. 1981. Epitaph to a scientific gadfly. (C. D. Darlington, FRS, the renowned cytogeneticist, died on 26 March 1981.) Scientists Remember, *News and Views, South African Journal of Science* 77:195–96.

When Cyril Darlington died, Graham Baker asked me to record my recollections of him in a contribution to Scientists Remember. *The idea appealed to me, and I went off and wrote it over the weekend.*

I had met Darlington while I was at Oxford and when he visited South Africa in 1963. Although differing from him on many issues, scientific and general, I still admired him as a man of independent mind and a forthright manner and as a pioneer cytogeneticist. I feel that my portrait explains clearly what it was about Darlington that appealed to me and also my reservations about him as a scientist and a man.

I should say something about one of his controversial aspects—his interest in racial genetics. I did not wish to deal with this directly, but wished to give some perspective to the matter, and so I chose a quotation of Darlington's referring to the Lysenko affair in which he had made a damning reference to the assault on truth and reason in science, both in Stalin's Russia and Hitler's Germany. I felt that this choice of quotation allowed me to illustrate the complexity of his views and to demonstrate how over-simplistic were some of the assessments of Darlington.

Darlington was very much in my mind at the time of his death because I had just completed a review of M.J.D. White's Modes of Speciation, *which was stimulated in part by the spate of uncritical reviews the book had received from the leading evolutionists of the day. How different were these uncritical reviews from the review by Darlington of the second edition of White's* Animal Cytology and Evolution, *in* Nature *on New Year's Day, 1955. This was considered so unfairly critical by White's scientific friends that they responded by sending their views to the journal. But there was more to Darlington's review than mere spleen. His book*

*reviews were never the meaningless, bland affairs that fail to
inform the potential readership that we have now come to
expect as the norm. They at least revealed to the reader
exactly what Darlington thought.*

*Looking back, I now see that my sketch of Darlington was
incomplete without mention of L. F. La Cour.*

All eukaryote geneticists and cytogeneticists know about Darlington,
"whose industry, insight and imagination transformed the study of
chromosomes from a crude science to a fine art."[1] Although his more
famous books[2-4] are still read, his essays and book reviews are now
seldom referred to. Rather than listing his numerous technical
achievements, I propose attempting to invoke the critical spirit of this
gadfly with a Northerner's distaste for the meretricious and pompous
in science.

I first turned to an essay which had stimulated me as an under-
graduate, "The Dead Hand on Discovery."[5,6] After thirty-three years,
I found that I still respond to its heady iconoclasm, and rereading it
has led to my discovering that much of my philosophy stems from
the mind of Darlington, its author! He would have been pleased to
hear the crows of delight which issued from the Hutchinson Laboratory as the postgraduates recognized that the fingerprints and
thumbprints of the dead hand are still all about us. Scientific discovery
"is a challenge and a threat. It challenges the respect due to parents,
to institutions and to governments. It threatens the authority of scripture, the sanctity of marriage and the sovereignty of nations. It dis-
turbs practices sanctified by generations of tradition, equally in the
farmyard, in the school and in the council chamber. It uproots belief.
It destroys property. It endangers life," or so the public thinks, according to Darlington.[5] "At bottom the conflict is not between scientist
and layman . . . It is between the rival powers of tradition, belief and
security—represented by all the established organs of society—on the
one hand; and innovation, doubt and discovery, on the other. It is
between those who have large fixed interests of the mind and those
who have not." Stirring words! And the universities did not escape:
"When we come to a teaching organization like a university we find
that it presents far greater obstacles to innovation than a mere academy. . . . There are three steps by which new knowledge is resisted in
our universities. First, the subject is kept out of the curriculum as
being debatable, premature, abstruse, or superfluous: too difficult,
the professors say for the undergraduates; too difficult, the under-

graduates know for the professors." And so on. What has changed in thirty years?

In the same essay Darlington attacked the compartmentalization of knowledge in universities: "The first powerful blows which might have broken down the barriers between the compartments of human knowledge were made by the work of Darwin. But Darwin's teaching falls in no compartment. It cannot therefore be taught, and it has in fact never seriously penetrated into our universities. Nearly 100 years after the publication of his great book, the *Origin of Species*, it is still possible to take degrees in Botany, Zoology or Medicine in any British university without any understanding of the work of Darwin." Within ten years, Darlington's advocacy of Darwin had become muted.[7] Darlington then provided bibliographic evidence which suggested that Darwin was subject to the "vice" of plagiarism. It was characteristic of the gadfly that this criticism should have appeared in the centenary year of the publication of the *Origin of Species*, a time when most biologists were eulogizing Darwin's genius!

Forthright criticism of books is not much in evidence today. It seems to be inhibited by some unwritten rules which depict such honesty as, somehow, unsporting. Book reviews have become a less reliable source of information on book quality, so much so that some books achieve unmerited circulation, to the detriment of learning. Again, in this regard, we miss Darlington. Sometimes his reviews were, arguably, immoderate. His review in *Nature*[8] of the second edition of M.J.D. White's book, *Animal Cytology and Evolution*, might be cited as an example. However, even in such cases his criticisms had a basis in great personal experience and should not be regarded as merely idiosyncratic. An example of a review which he wrote almost a quarter of a century later demonstrates that he remained a severe critic till late in life, perhaps to the end. The review is of G. E. Allen's *Thomas Hunt Morgan*.[9] "After the lapse of years Morgan, having lost interest in the experiments of genetics, and never having sustained an interest in its ideas, was awarded a Nobel Prize. Dr Allen's account allows us a glimpse of the embarrassment that once again overtook him. Having delayed his journey to Stockholm, on his way he arrived in Edinburgh still uncertain what, and whom, and how much, he should acknowledge." Toward the end of the review Darlington turned his attention to the unfortunate author: "Dr Allen must have occupied many years in preparing this work, as his method of assembling it is a recipe for chaos, chaos in writing and chaos in reading. His accommodation of Bateson (and the present reviewer) at the John Innes Horticultural 'School' which he places in Merton College, Cambridge, is an extreme example." Readers of such reviews are not only entertained, but also instructed, and they gained a clear appreciation of the book's worth. What more can one expect from a review?

Darlington served genetics well through his critical writings. It, therefore, comes as something of a shock to notice how often his critical faculties deserted him in his own writings, particularly those on evolution. His style in these works is characterized by assertiveness, poor documentation, and a lack of attention to the fine details of causality. This latter point may be illustrated here by considering Darlington's treatment of the phenomenon of incompatibility in flowering plants. Darlington regards self-incompatibility as an outbreeding system in flowering plants. How can such a view be justified? How could it evolve under natural selection? Of course it can't, any more than interspecific sterility can; and it is surprising that a scientist of Darlington's standing should think it could.

Sometimes the very exuberance of his style led to ambiguity. Just what does he mean when he writes[3] that "the species is prepared by its genetic system for what we may call unexpected events. Natural selection has provided it with a system which although automatic is not properly described [p. 196] as blind. On the contrary it has been endowed with an unparalleled gift, an automatic property of foresight." As with many other cytogeneticists and geneticists, Darlington seems never to have grasped the unique features of species and speciation,[3,10,11] and his later views were essentially similar to his earlier. This, of course, did not stop him writing on the subject with his usual confidence! I shall illustrate his proneness to logical lapses with two sentences which lie a page apart:[10] "In other groups . . . tetraploid forms and the diploids from which they have been derived have been assigned, as they should be assigned, to different species." He then wrote, "It thus happens that within the same species *Allium schoenoprasum* L. . . . we may have both types of tetraploid, one non-hybrid and fairly fertile [autotetraploid] and the other hybrid and highly fertile [allotetraploid]; probably combinations of the two occur." Again, like many other cytogeneticists, he demonstrated a distinct tendency to accept sympatric speciation mediated by chromosomal rearrangements.

The dismal history of genetics in the Soviet Union, culminating in the Lysenko affair, stimulated some of Darlington's most fearless criticism.[12] This criticism tells us much about the man who made it, and we need periodic reminding of this example of doctrinal interference with science, and so my last quotation will be somewhat longer:

The Russian story is . . . one not to be remedied merely by the exposure of a few charlatans. The leading Russian geneticists (apart from those who have taken refuge outside Soviet-controlled countries) have been "liquidated" in the course of this long political intrigue. These are no longer questions that can be argued about in Russia. They have been decided. All those who were prepared to argue have been put away . . .

We see indeed the official overthrow of truth and reason and of the men who stood by them in one branch of science. This overthrow is no less official than that in Hitler's Germany. It affects first and foremost the same branch of science, that concerned with heredity, race, and class. The reason is merely that this is the science of greatest political moment under the conditions of social stress in the world today. The implications of this situation reach beyond the scope of the present review. But men of science everywhere will do well to ponder them.

I did not know Darlington well, though I met him quite often at Oxford and then again in South Africa in 1963. To a young "unknown" he was patient and kind and encouraging. I did not know him well enough to be able to predict how he would have stood the little bit of his own medicine I have administered here. I like to think he would have enjoyed its contrivance! In any case, I shall, personally, never be free of his influence. We need more gadflies.

REFERENCES

1. Lewis, K. R., and B. John. 1963. *Chromosome Markers*. London: Churchill.
2. Darlington, C. D. 1937. *Recent Advances in Cytology*. 2d ed. London: Churchill.
3. Darlington, C. D. 1958. *Evolution of Genetic Systems*. 2d ed. Edinburgh: Oliver and Boyd.
4. Darlington, C. D., and L. F. La Cour. 1942. *The Handling of Chromosomes*. London: Allen and Unwin.
5. Darlington, C. D. 1948. The dead hand on discovery, I. The fingers of learning. *Discovery* 9:358–62.
6. Darlington, C. D. 1949. The dead hand on discovery, II. The thumb of office. *Discovery* 10:7–11.
7. Darlington, C. D. 1959. *Darwin's Place in History*. Oxford: Blackwell.
8. Darlington, C. D. 1955. Genetics and the chromosomes. *Nature* 175:4–5.
9. Darlington, C. D. 1979. Morgan's crisis. *Nature* 278:786–87.
10. Darlington, C. D. 1940. Taxonomic Species and Genetic Systems. Chap. 4 in *The New Systematics*, ed. J. S. Huxley, 137–60. London: Oxford University Press.
11. Darlington, C. D. 1978. A diagram of evolution. *Nature* 276:447–52.
12. Darlington, C. D. 1947. A revolution in Soviet science. *Discovery* 8: 40–43.

7 Morphological Resemblance and Its Relationship to Genetic Distance Measures

Lambert, D. M., and H.E.H. Paterson. 1982. Morphological resemblance and its relationship to genetic distance measures. *Evolutionary Theory* 5:291–300.

This paper launched Dave Lambert's career as a conference speaker. He wrote this paper, but it included ideas from both of us. It showed the value of the recognition concept in evolutionary explanation and effectively countered proposals, which were common at that time, that evolutionary status could be assigned to diverged populations in terms of identity- or difference-indices. There is more to this matter than we covered in this paper, and the subject requires deeper exploration. The paper also contains some additional detail on the stability of the specific-mate recognition system to that provided in Paterson (1978) (chap. 3).

ABSTRACT

The advent of a number of genetic and cytogenetic approaches to the measurement of the distance between populations and species has radically changed the study of the relationships between groups of animals. Morphological resemblance, as a measure of divergence, can now be compared with results obtained using these new techniques. This contribution is an attempt to consider the degree of concordance of these different approaches. Particular attention is paid to species of the genus *Drosophila*; however, a broad range of groups is discussed as well. It is argued that it is not meaningful to relate any genetic divergence measure to the taxonomic categories: subspecies, "sibling" species, and "nonsibling" species. With genetical species no relationship exists between morphological divergence and the two most common forms of genetic distance, chromosomal and electromorphic distance. If speciation is thought of in terms of a change from one specific-mate recognition system (Paterson, 1976 [chap. 2], 1978 [chap. 3], 1980 [chap. 4], 1981 [chap. 5]) to another, one would not expect any consistent relationship to exist.

There has been a great deal of discussion in recent times regarding the genetic divergence which is to be expected between groups of organisms whether these groups be species, subspecies, etc. The question is an important one since these genetic distances between groups have been used to construct dendrograms of relatedness. Inherent in many discussions is a belief that speciation events are accompanied by a set amount of genetic change (e.g., Avise, 1976; Ayala, 1975). This assumption is vital if degrees of relatedness are to be assessed using the genetic distance method. If speciation events are not accompanied by a set amount of divergence, as measured by the two most common estimates of genetic distance, chromosomal and electromorphic distance, then the method must be treated with extreme caution.

Any discussion of speciation should be based on a firm understanding of the nature of species. This must surely be the logical beginning (Paterson, 1981 [chap. 5]).

SPECIES CONCEPTS

The most commonly accepted species definition is one based on reproductive isolation. Species are, according to this view, "groups of organisms which are reproductively isolated from other groups" (Mayr, 1963). In keeping with this viewpoint, Dobzhansky (1970) commented, "Species are . . . systems of populations; the gene exchange between systems is limited or prevented in nature by a reproductive isolating mechanism or perhaps by a combination of several such mechanisms" (Dobzhansky, 1970:357). This has been called the isolation concept (Paterson, 1978 [chap. 3]). Isolating mechanisms are generally regarded as comprising two broad classes: premating and postmating mechanisms (Mayr, 1963; Dobzhansky, 1970; White, 1978; Stebbins, 1950; Grant, 1971). [p. 292]

Since, however, postmating isolating mechanisms cannot be directly selected for, as realized originally by Darwin (1859), then these are surely not mechanisms which evolved for the function of isolation (*sensu*, Williams, 1966). Similarly, as Paterson (1981 [chap. 5]) has emphasized, many authors such as White (1978), Dobzhansky (1970), and Mayr (1963) have argued that there is indisputable evidence that species arise in allopatry. Clearly then, species that arise in such a way should not be described as having acquired "isolating mechanisms" since these have evolved in geographic separation from the parental population.

Dobzhansky (1951, 1970) maintained logical consistency by advocating a model of speciation which allowed a direct role for natural selection in the origin of species. This is the model of speciation by reinforcement. However, Paterson (1978 [chap. 3]) and Moore (1957) have raised difficulties with the theoretical basis of this model. An increasing number of authors have also commented on the lack of convincing cases in the literature (Mayr, 1963; Loftus-Hills, 1975; Jackson, 1973; Futuyma and Mayer, 1980).

The recognition concept defines species purely in relation to conspecifics. According to this view species are groups of organisms which are tied together by a common communication system. This has been called the specific-mate recognition system. Mayr (1963) views species in both negative ("isolation") and positive ("recognition") terms; e.g., "The ethological isolating mechanisms then, are barriers to mating due to incompatibilities in behavior [negative isolation aspect]. The males of every species have specific courtships or displays to which, on the whole, only females of the same species are receptive [positive or recognition aspect]" (Mayr, 1963:95). The recognition concept views "postmating isolating mechanisms" as unsatisfactory criteria for delineating species (Paterson, 1981 [chap. 5]). Postmating effects are instead considered as intraspecific phenomena. This is reasonable since in angiosperms incompatibility genes are easily accommodated in this way. Similarly, in some animals, such as the mosquito species *Culex molestus*, sterility is a within-species phenomenon (see Paterson, 1981, [chap. 5] for an extensive discussion of these points).

If the SMRS of individuals of a daughter population changes such that they no longer recognize as mates members of the parental population, then a speciation event has occurred. This incidentally also results in the origin of a new and distinct gene pool.

SPECIES AND THEIR CHARACTERISTIC STABILITY

Since the recognition concept solves many of the conceptual difficulties inherent in an isolationist approach to the study of species, this discussion will be framed in terms of species as a group of individuals which share a common SMRS. This is a genetically programmed system which has been acted upon by natural selection to function efficiently in the preferred habitat of a species and results in conspecific syngamy. Paterson (1976 [chap. 2]) has discussed details of the system and pointed out that it is under stabilizing selection. This is because individuals which are deviant in one or more of the steps comprising the SMRS are less likely to be recognized as mates by conspecifics. The more deviant an individual is, the lower is its selective advantage. We would then expect that the SMRS will show a stability within a species that is not shown by most other characters, as was realized by Walker (1964:351–53).

Recent studies on *Drosophila melanogaster* provide evidence in support of this expectation. Teissier (1958) and Takanashi and Kitagawa (1977) have shown that French and Japanese populations of this species differ considerably as reflected by biometric techniques. Kojima et al. (1970) and Girard and Palabost (1976) show that significant differences in the frequencies of electromorphs exist between the two populations. However, choice experiments between these two populations, reported by Petit et al. (1976), reveal that the SMRSs of the two populations have not diverged [p. 293] and that individuals of either population do not distinguish between Japanese and French mates. David and Bocquet (1975) and David et al. (1977) have compared biometrically African and French strains of *Drosophila melanogaster* and found them to be very divergent. In fact the populations have non-overlapping distributions in a bivariate plot of characters such as female ovariole number and female abdominal chaetae length. Tests in progress, designed to compare the degree of divergence of the SMRSs of the populations, have not revealed any divergence (David, personal communication). A similar case illustrating a lack of variation in the SMRS is that of Anderson and Ehrman (1967). These authors demonstrated random mating between widespread populations of *Drosophila pseudoobscura* from North America.

These studies provide evidence for the geographic stability of the SMRS. This is vital so that individuals of any sexually reproducing species can recognize conspecific mates for the purpose of sexual reproduction. Although other coadapted characters exist, many which are not involved in mate recognition are not subject to these same constraints and hence are more likely to respond to local selective pressures and show greater variation.

If each biparental species has its own SMRS and hybrids are rare in nature, then speciation occurs when a daughter population moves from the SMRS of a parental population to another quite distinct and effectively nonoverlapping system. This involves a change in the male-female communication system of the original population from which the founder came and the restabilization of the new system in the daughter species. When the population increases to a large size, the SMRS becomes stabilized and hence is buffered from further change. Only under conditions of small population size is it likely that the SMRS can change and again become fixed. Sometimes in small peripheral isolates natural selection can directly cause a change in the SMRS resulting in an increase in its efficiency in the new environment. The alternative, another likely outcome of such a situation, is that the population will become extinct. Another possible process that can lead to a new SMRS is that in a small isolate the SMRS of one sex could be pleiotropically affected by another gene. If this gene confers a selective advantage upon individuals carrying it, and the population is small, it may become fixed. Under this no-choice situation through many generations, the SMRS of the unaffected sex will be selected to recognize once again the signals of the other sex. A good model for such a process is the *yellow* allele in *Drosophila melanogaster* (Bastock, 1956). *Yellow* males have a deviant courtship pattern while *yellow* females are normal. Since the gene also confers on its carrier a resistance to starvation (Kalmus, 1941) this could, in a certain new environment, lead to its fixation in a small isolate. Females would then be selected for improvement in recognition of the mutant male SMRS.

The work of David, Teissier, and Petit et al., presented earlier, shows that morphologically and electromorphically populations of a species can diverge, while the system of mate recognition remains stable. In a large population the recognition system should be resistant to change owing to the difficulty in fixing genes in anything but small populations.

EVOLUTIONARY BIOLOGY'S "CARDINAL PROBLEM"

A genetic consequence of this concept of a species is that after any particular speciation event the degree of divergence in genes controlling external morphological characters is not necessarily representative of the total of all the genetic changes that have occurred. If a speciation event is the change from one stabilized SMRS to another reconstructed and restabilized SMRS, then this will sometimes be accompanied by very small changes in the total genome and sometimes by larger changes. Ayala (1975) has described the question of the amount of genetic change that occurs at speciation as "the cardinal problem of evolutionary biology." Nevo and Cleve (1978) have recently said that "How many and what kind of genes are implicated in speciation? [p. 294] is a central unresolved problem of evolutionary biology. Does speciation require major genomic changes or may minor ones suffice?" Lewontin in his influential book (1974:186) has suggested that the genetic differences between sibling species will always be greater than 10 percent of the total genome as reflected by electrophoresis studies.

In keeping with this inherent supposition that a speciation event is marked by a set amount of genetic divergence Ayala (1975) and Avise (1976) have maintained that a certain degree of genetic change can be equated to various taxonomic categories, e.g., subspecies, "sibling" species. If speciation is thought of in terms of a change in SMRS then one would not expect any consistent relationship to exist. It is argued here that a thorough search of the literature reveals that this expectation is verified.

A CONSISTENT RELATIONSHIP?—EVIDENCE FROM THE LITERATURE

The important Hawaiian *Drosophila* project (see Carson et al., 1970, for review) has provided a number of evolutionary examples which have a bearing on this general question. Firstly, *D. heteroneura* and *D. silvestris* (of the *planitibia* subgroup) are two closely related species which are sympatric at many localities on the island of Hawaii. These species are homosequential for the banding pattern on larval polytene chromosomes and have a genetic similarity coefficient of 0.95. On the scale provided by Ayala (1975) this is similar to the level of subpopulations of one species. However, morphologically, males of these species are very distinct (see Val, 1977, for details) and were originally (i.e., before the use of these genetic techniques) considered by systematists not to be closely related.

Another two Hawaiian *Drosophila* species, *D. setosimentum* and *D. ochrobasis*, members of the *adiastola* subgroup, are two closely related species and chromosomally divergent to the extent that they have a chromosomal similarity coefficient of 0.66–0.68 (Carson et al., 1975). The species have an electromorphic similarity coefficient, in some areas, of 0.98 (Carson et al., 1975). Again this falls within the range of populations of the one species according to Ayala (1975). Also within the *planitibia* subgroup *D. cyrtoloma*, *D. ingens*, and *D. melanocephala* have similarity coefficients of from 0.99 to 0.96 (Johnson et al., 1975). These examples alone demonstrate the lack of utility in Ayala and co-workers' proposal.

Cases such as these are readily understood when speciation is thought of in terms of changes in the SMRS. Many Hawaiian *Drosophila* perform lek mating behavior. Males attract females to a courtship site away from the feeding area. They display to females, and these males' signals, such as the waving of patterned wings and the presence of elaborate head characteristics, are certainly involved in the SMRS (Spieth, 1974). These species use optical signals in mate recognition, and hence we can also use them in order to recognize these discrete units in nature.

A number of the species of the *Anopheles gambiae* complex of mosquitoes are morphologically identical despite many attempts to distinguish them (Coluzzi, 1964; Green, 1972). These species have diverged enzymatically, to the extent that Mahon et al. (1976) and Miles (1978) have used electrophoresis as a reliable method of identification of these cryptic species. Presumably males and females of these species use nonmorphological cues in their recognition of conspecific mates (i.e., in the SMRS), and the degree of morphological change associated with a change in the SMRS in this group in no way reflects the degree of total genetic change that has occurred.

Indeed, the mere existence of cryptic species, with varying degrees of genetic similarity, argues against any consistent relationship between morphological and genetic change at speciation. It might be suggested that these varying degrees of genetic change are the result of different types of speciation. However, we regard nonallopatric models of speciation as unconvincing (as do Futuyma and Mayer [1980]). [p. 295] Chromosomal changes have not been shown to cause speciation, for example; they have only been shown to be correlated with it. If speciation occurs via small, isolated populations, this would be expected. Paterson (1981 [chap. 5]) has discussed a number of these models and concluded that convincing evidence is available for speciation of allopatric populations only. Speciation via founder populations would indeed predict varying degrees of genetic change at speciation.

Lambert (1979, 1981) has presented cytogenetic evidence for the existence of four cryptic species within the African mosquito "*Anopheles marshallii.*" Ovarian polytene chromosome analysis shows a total lack of interbreeding between individuals of these species in the wild, even though they are sympatric over wide areas. This indicates that complete divergence in the SMRSs of these species has occurred. The species are morphologically very alike and adults can only be separated by multivariate analysis of wing and palp characters. These species are electromorphically very similar, having similarity coefficients of from 0.9718 to 0.8027 (Lambert, 1983).

Although they do not commit themselves with regard to the status of their forms, Cardé et al. (1978) have presented very good evidence of the existence of two previously unsuspected cryptic species within the European corn borer (*Ostrinia nubilalis*). The two species use the same chemical components, Z-11 and E-11 tetradecenyl acetate, in their pheromones. These are involved in mate recognition. However, they are present in different proportions—one uses a mixture of 97 : 3 and the other 4 : 96. The two species are very similar as indicated by electromorph studies, having a similarity coefficient of 0.997.

Field trapping experiments by Cardé et al. (1978) illustrate the individuals of the two species respond only to the blend characteristic of their own species. In this case a change in the SMRS has been accompanied by very little electromorphic divergence.

Any estimation of the degree of genetic divergence which accompanies a change in the SMRS must be complicated by the factor of postspeciational change. However, data such as those discussed for Hawaiian *Drosophila*, "*Anopheles marshallii*," and the European corn borer indicate that some speciation events, at least, are accompanied by very little detectable change. It is certainly at least possible that regulatory genes are involved in the genetic changes associated with speciation (Wilson, 1975). This would then account for the low levels of divergence as reflected in electromorph studies. At this stage, however, we consider that the evidence for this is suggestive but certainly not conclusive.

In animals which use nonmorphological cues in the recognition of conspecific mates, speciation can occur without any accompanying divergence in morphology. If the divergence in genes controlling external morphology is no reflection of the total genetic changes that have occurred in the speciation events, then we should not be surprised to find two such strikingly different degrees of electromorphic divergence as that found in the corn borer and the *A. gambiae* group of species. It is conceivable that very small changes might accompany the change in pheromone components in the two species of corn borers, for example, and perhaps much more considerable changes in the case of some of the *A. gambiae* species. This may well be related to the rapidity with which each individual speciation event occurs. If it happens quickly, and the population size increases quickly, few genetic differences may result. However, if it takes a longer period, more changes are likely to accumulate. A flush-crash cycle such as detailed by Hampton Carson (1975) could conceivably result in an altered SMRS becoming fixed in an allopatric isolate of a parental species while resulting in only minor electromorphic divergence. See Powell (1978) for some experimental support for Carson's theory. Similarly, the degree of genetic change may be dependent upon the characteristics of the environment in which the small population is restricted. Drastic changes in characteristics of the available habitat may, for example, result in substantial genetic change accompanying speciation. [p. 296]

Maxson and Wilson (1974) have presented a case of the two North American tree frogs (*Hyla eximia* and *Hyla regilla*) which are morphologically almost identical and yet vary considerably on proteins, so much so that the species differ as much from each other as they do from species outside the genus *Hyla*. Perhaps the best example of small genetic divergence accompanied by very considerable morphological changes is the case of man and chimpanzees (King and Wilson, 1975). These species have a similarity coefficient which is equal to that found for many cryptic species and yet are morphologically quite distinct. Recently Bruce and Ayala (1978) have reported the results of electromorphoresis studies which show that human proteins are very similar to the proteins of seven species of apes studied.

Lyons et al. (1977, 1980) have recently shown the existence of two biologically distinct species with different chromosome numbers, $2n = 32$ and $2n = 36$, within the taxonomic species *Mastomys natalensis*. This "species" has been implicated as a carrier of Lassa virus and Plague bacilli in Africa. These studies have shown that there is little homology between the G-banding patterns of the species. Further studies (Green et al., 1978) have revealed a divergence in hemoglobin patterns of the two species as shown by gel electrophoresis. Consequent attempts to distinguish the two species using discriminant function analysis on morphological characters have yielded only limited results (Gordon, personal communication), indicating the high degree of morphological similarity of the species. Mate recognition in these species is achieved at least partly by ultrasonic communication, and the two species have diverged considerably in this regard (Gordon, personal communication).

Since hemoglobins are such evolutionarily stable molecules (Dickerson and Geis, 1969) and considering the results of the G-banding studies, these species appear to have been separate for a considerable period of time, yet morphologically they are extremely similar. Obviously a phenetic approach to such a problem is completely inappropriate, as Green et al. (1978) have stressed. Only a genetical or behavioral approach to such a situation could have revealed the existence of these two cryptic species.

A number of cases which have been proposed as examples of divergence with little protein differentiation have appeared in the literature (Avise et al., 1975; Gartside et al., 1977; Nevo and Cleve, 1978; Nevo and Shaw, 1972; Prakash, 1972, 1977; Turner, 1974; Brown, 1981). The degrees of morphological divergence which have accompanied these changes vary from quite considerable, in the case of the Californian minnows, to almost none in the case of the species of the complex of mole rats and the ribbon snakes.

CONCLUSION

At the level of closely related species the argument presented here is that there is no consistent correlation between morphological resemblance and genetic distance, because no set amount of genetic divergence can be found to accompany speciation events. Similarly, depending upon the predominant communication channel (visual, chemical, auditory, etc.) that the species use in mate recognition, there will be varying degrees of morphological divergence associated with a change in the SMRS. It should be pointed out that, in the long term, the longer the time since the divergence of two species from a common ancestral population, the greater the genetic distance can be expected to be.

ACKNOWLEDGMENTS

We thank the following colleagues for reading the manuscript: Elisabeth Vrba, Chris Green, Marc Centner, Judy Masters, Brian Levey, Linda Maxson, and Amanda Harper. We are grateful also to Eviatar Nevo and an anonymous referee for their useful comments on the work.

REFERENCES

Anderson, W. W., and C. Ehrman. 1967. Mating choice in crosses between geographic populations of *Drosophila pseudoobscura*. *American Midland Naturalist* 81:47–53.

Avise, J. C. 1976. Genetic Differentiation during Speciation from Molecular Evolution. In *Molecular Evolution*, ed. F. J. Ayala, 106–22. Sunderland, Mass.: Sinauer Associates, Inc.

Avise, J. C., J. J. Smith, and F. J. Ayala. 1975. Adaptive differentiation with little genic change between two native Californian minnows. *Evolution* 29:411–26.

Ayala, F. J. 1975. Genetic differentiation during the speciation process. *Evolutionary Biology* 8:1–78.

Bastock, M. 1956. A gene mutation which changes a behavior pattern. *Evolution* 10:421–39.

Brown, K. 1981. Low genetic variability and high similarities in the crayfish genera *Cambarus* and *Procambarus*. *American Midland Naturalist* 105:225–32.

Bruce, E. J., and F. J. Ayala. 1978. Humans and apes are genetically very similar. *Nature* 276:264–65.

Cardé, R. T., W. L. Roelofs, R. G. Harrison, A. T. Vawter, P. F. Brussard, A. Mutuura, and E. Munroe. 1978. European corn borer: Pheromone polymorphism or sibling species? *Science* 199:555–56.

Carson, H. L. 1975. The genetics of speciation at the diploid level. *American Naturalist* 109:83–92.

Carson, H. L., D. Elmo Hardy, H. T. Spieth, and W. S. Stone. 1970. The Evolutionary Biology of the Hawaiian Drosophilidae. In *Essays in Evolution and Genetics in Honor of Th. Dobzhansky,* ed. M. K. Hecht and W. C. Steere, 437–544. Amsterdam: North-Holland Publishing Company.

Carson, H. L., W. E. Johnson, P. S. Nair, and F. M. Sene. 1975. Allozymic and chromosomal similarity in two *Drosophila* species. *Proceedings of the National Academy of Science,* USA 72:4521–25.

Coluzzi, M. 1964. Morphological divergences in the *Anopheles gambiae* complex. *Rivista di Malariologia* 43:197–232.

Darwin, C. 1859. *On the Origin of Species by Means of Natural Selection; or, the Preservation of Favoured Races in the Struggle for Life.* London: John Murray.

David, J. R., and C. Bocquet. 1975. Evolution in a cosmopolitan species: Genetic latitudinal clines in *Drosophila melanogaster* wild populations. *Experientia* 31:164–66.

David, J. R., C. Bocquet, and M. de Scheemaeker-Louis. 1977. Genetic latitudinal adaptation of *Drosophila melanogaster*: New discriminative biometrical traits between European and equatorial African populations. *Genetical Research, Cambridge* 30:247–55.

Dickerson, R. E., and I. Geis. 1969. *The Structure and Action of Proteins.* New York: Harper and Row.

Dobzhansky, T. 1951. *Genetics and the Origin of Species.* 3d ed. New York: Columbia University Press.

———. 1970. *Genetics of the Evolutionary Process.* New York: Columbia University Press.

Futuyma, D. J., and G. C. Mayer. 1980. Non-allopatric speciation in animals. *Systematic Zoology* 29:254–71.

Gartside, D. F., J. S. Rogers, and H. C. Dessauer. 1977. Speciation with little genic and morphological differentiation in the ribbon snakes *Thamnophis proximus* and *T. sauritus* (Colubridae). *Copeia* 4:697–705.

Girard, P., and L. Palabost. 1976. Etude du polymorphisme enzymatique de populations naturelles françaises de *Drosophila melanogaster. Archives de Zoologie expérimentale et générale* 117:41–55.

Grant, V. 1971. *Plant Speciation.* New York: Columbia University Press.

Green, C. A. 1972. The practical problem of identifying members of the *Anopheles gambiae* complex in autecological studies. *Parasitologia* 13:421–27.

Green, C. A., D. H. Gordon, and N. F. Lyons. 1978. Biological species in *Praomys (Mastomys) natalensis* (Smith), a rodent carrier of Lassa virus and bubonic plague in Africa. *American Journal of Tropical Medicine and Hygiene* 27:627–29.

Jackson, R. C. 1973. Chromosomal evolution in *Haplopappus gracilis*: A centric transposition race. *Evolution* 27:243–56.

Johnson, W. E., H. L. Carson, K. Y. Kaneshiro, W.W.M. Steiner, and M. M. Cooper. 1975. Genetic Variation in Hawaiian *Drosophila*, II. Allozymic

Differentiation in the *D. planitibia* Subgroup. In *Isozymes. Genetics and Evolution*, ed. C. L. Markert, 563–84. Vol. 4. New York: Academic Press.

Kalmus, H. 1941. The resistance to desiccation of *Drosophila* mutants affecting body colour. *Proceedings of the Royal Society* 130:185–201.

King, M. C., and A. C. Wilson. 1975. Evolution at two levels in humans and chimpanzees. *Science* 188:107–16.

Kojima, K. I., J. Gillespie, and Y. N. Tobari. 1970. A profile of *Drosophila* species enzymes assayed by electrophoresis. *Biochemical Genetics* 4:627–37.

Lambert, D. M. 1979. *Anopheles marshallii* (Theobald) is a complex of species. *Mosquito Systematics* 11:173–78.

———. 1981. Cytogenetic evidence of a possible fourth cryptic species within the taxon *Anopheles marshallii* (Theobald) (Diptera: Culicidae) from Northern Natal. *Mosquito Systematics* 13:168–75.

———. 1983. A population genetical study of the African mosquito *Anopheles marshallii* (Theobald). *Evolution* 37:484–95.

Lewontin, R. C. 1974. *The Genetic Basis of Evolutionary Change*. New York: Columbia University Press.

Loftus-Hills, J. J. 1975. The evidence for reproductive character displacement between the toads *Bufo americanus* and *B. woodhousii fowleri*. *Evolution* 29:368–69.

Lyons, N. F., D. H. Gordon, and C. A. Green. 1980. G-banding chromosome analysis of species A of the *Mastomys natalensis* (Smith, 1834) (Rodentia, Muridae) from Rhodesia. *Genetica* 54:209–12.

Lyons, N. F., C. A. Green, D. H. Gordon, and C. R. Walters. 1977. G-banding chromosome analysis of *Praomys natalensis* (Smith) (Rodentia: Muridae) from Rhodesia, I. 36 Chromosome population. *Heredity* 38:197–200.

Mahon, R. J., C. A. Green, and R. H. Hunt. 1976. Diagnostic allozymes for routine identifications of adults of the *Anopheles gambiae* complex (Diptera, Culicidae). *Bulletin of Entomological Research* 66:25–31.

Maxson, L. R., and A. C. Wilson. 1974. Convergent morphological evolution detected by studying proteins of tree frogs in the *Hyla eximia* species group. *Science* 185:66–68.

Mayr, E. 1963. *Animal Species and Evolution*. Cambridge, Mass.: Belknap Press of Harvard University Press.

Miles, S. J. 1978. Enzyme variation in the *Anopheles gambiae* Giles group of species (Diptera: Culicidae). *Bulletin of Entomological Research* 68:85–96.

Moore, J. A. 1957. An Embryologist's View of the Species Concept. In *The Species Problem*, ed. E. Mayr, 325–38. Publication No. 50. Washington, D.C.: American Association for the Advancement of Science.

Nevo, E., and H. Cleve. 1978. Genetic differentiation during speciation. *Nature* 275:125–26.

Nevo, E., and R. C. Shaw. 1972. Genetic variation in a subterranean mammal *Spalax ehrenbergi*. *Biochemical Genetics* 7:235–41.

Paterson, H.E.H. 1976. The role of postmating isolation in evolution. Invited Lecture Fifteenth International Congress of Entomology, Washington, D.C., Symposium on the "Application of Genetics to the Analyses of Species Differences." Unpublished MS. [This volume, chap. 2.]

————. 1978. More evidence against speciation by reinforcement. *South African Journal of Science* 74:369–71. [This volume, chap. 3.]

————. 1980. A comment on "Mate Recognition Systems." *Evolution* 34:330–31. [This volume, chap. 4.]

————. 1981. The continuing search for the unknown and unknowable: A critique of contemporary ideas on speciation. *South African Journal of Science* 77:113–19. [This volume, chap. 5.]

Petit, C., O. Kitagawa, and T. Takamura. 1976. Mating system between Japanese and French geographic strains of *Drosophila melanogaster*. *Japanese Journal of Genetics* 51 (2):99–108.

Powell, J. R. 1978. The founder-flush speciation theory: An experimental approach. *Evolution* 32:465–74.

Prakash, S. 1972. Origin of reproductive isolation in the absence of apparent genic differentiation in a geographic isolate of *Drosophila pseudoobscura*. *Genetics* 72:143–55.

————. 1977. Further studies on gene polymorphism in the mainbody and geographically isolated populations of *Drosophila pseudoobscura*. *Genetics* 85:713–19.

Spieth, H. T. 1974. Mating Behavior and Evolution of the Hawaiian *Drosophila*. In *Genetic Mechanisms of Speciation in Insects*, ed. M.J.D. White, 94–101. Sydney: Australia and New Zealand Book Company.

Stebbins, G. L. 1950. *Variation and Evolution in Plants*. New York: Columbia University Press.

Takanashi, E., and O. Kitagawa. 1977. Quantitative analysis of genetic differentiation among geographical strains of *D. melanogaster*. *Drosophila Information Service* 52:28.

Teissier, G. 1958. Distinction biometrique des *Drosophila melanogaster* françaises et japonaises. *Annales de génétique* 1:2–10.

Turner, B. J. 1974. Genetic divergence of Death Valley pupfish species: Biochemical versus morphological evidence. *Evolution* 28:281–94.

Val, F. C. 1977. Genetic analysis of the morphological differences between two interfertile species of Hawaiian *Drosophila*. *Evolution* 31:611–29.

Walker, T. J. 1964. Cryptic species among sound-producing ensiferan Orthoptera (Gryllidae and Tettigoniidae). *Quarterly Review of Biology* 39:345–55.

White, M.J.D. 1978. *Modes of Speciation*. San Francisco: W. H. Freeman.

Williams, G. C. 1966. *Adaptation and Natural Selection*. Princeton. N.J.: Princeton University Press.

Wilson, A. C. 1975. Evolutionary importance of gene regulation. *Stadler Genetics Symposium* 7:117–33.

NOTE

Based on a paper presented to the Zoological Society of Southern Africa Symposium on Systematics, Pretoria, 12–13 September 1978.

8 Perspective on Speciation by Reinforcement

Paterson, H.E.H. 1982. Perspective on speciation by reinforcement. *South African Journal of Science* 78:53–57.

I was prompted to write this paper because I felt that Littlejohn's (1981) paper was likely to be seriously misleading to anyone attempting to assess the recognition concept. Although I have not spent time on sexual selection, others have made claims about the way sexual selection relates to SMRS (Verrell, 1988; Eldredge, 1989).

It has long been evident to me that many authors use words loosely. I try hard to be consistent in my use of terms that I have introduced. It was therefore important to correct Littlejohn's use of "mate choice," which was the very term I was attempting to avoid when I used the term "mate recognition." Another term I avoid absolutely is "species recognition." The concept of species is a very human one, and so only a human can practice "species recognition." What an organism can recognize is a specific mate. I have, of course, carefully defined what I mean by "recognition."

The term "sexual selection" has confused people since publication of the Origin of Species *in 1859, though Darwin was fairly careful to say what he meant by it. Unfortunately, as with "natural selection," he used the term for two phenomena: intrasexual competition for mates and female mate choice. However, some authors use the term more broadly to mean "all mechanisms which cause deviations from panmixia" (Ehrman, 1972:106). As I have often stated, "mechanism" has a specific meaning, something like "parts that work together as in a machine to achieve an end," which is very constraining. One can point out that age is a relevant factor with respect to panmixia among sea lions and deer, for example, but would Ehrman regard it as a character evolved under sexual selection?*

Disentangling characters evolved under sexual selection from those which evolved as SMRS characters is not always easy. However, it is essential to do so if we are to understand how each character evolved. For example, SMRS characters have evolved in conspecific mating partners to serve functions connected with the achieving of fertilization (attraction of a partner from a distance, recognition of a potential mate, recognition of the egg by a sperm, etc.). Spurs in gallinaceous birds are found in males and appear to be used in intermale conflict. They appear to have evolved under sexual selection. But the matter is not straightforward. Sex-pads on the front feet of Xenopus laevis *males appear to have evolved to facilitate the gripping of the females by the male frogs during amplexus. They are thus part of the fertilization system and did not evolve under sexual selection. In Noble's (1936) experiment with flickers, the black line on the head of males serves both as a signal indicating to female flickers that the bearer is a male, and as a signal indicating maleness to other male flickers, leading to intermale conflict within territories. Perhaps such a character evolved as a part of the SMRS and then evolved, secondarily, to function in intermale conflict. The red chest feathers of the European robin occur in both sexes. An isolated bunch of these feathers will stimulate aggressive behavior in male robins.*

I objected to a number of other assertions that Littlejohn made and expressed my views on them. I took the opportunity of adding to the list of procedures for improving our evolutionary models, which I had recommended in my 1981 paper (chap. 5).

This paper, then, can be seen as an amplification of the 1978 paper (chap. 3). It was in this paper that I first drew attention to the fundamentally important work of Russell Coope of the University of Birmingham. I had known of his work since 1978 when, from the floor, during a lecture I gave at the Linnean Society in London, one of his graduate students had drawn my attention to it. Ever since then his studies had influenced me very significantly, and it surprises me that I had not mentioned him earlier.

Any attempt at replacing a widely held concept demands determination from its author. Inevitably, change will be resisted, and in the process the new idea will be misrepresented, whether through carelessness, incomprehension, or other reasons. The author, it seems to me, has a responsibility to defend his ideas from misrepresentation or attack, or his critics may through default gain an undeserved victory to the detriment of science. However, science will benefit most if the response is as dispassionate as possible and is aimed at further clarifying any points which appear to have been misunderstood.

In a recent paper[1] Littlejohn has questioned aspects of my[2] objections to the current view of species[3] and has seriously misrepresented my suggested alternative concept of species. This contribution is meant to defend and strengthen my earlier criticism of the idea that speciation may occur through the reinforcement of isolating mechanisms by natural selection and to clarify further the provisions of the recognition concept of species.[2,4]

GENERAL CONSIDERATIONS

The paper [chap. 3] referred to by Littlejohn comprised the substance of a short address which I gave in front of the Genetical Society at Leeds in January 1978. Its aims were made clear by its title: "More Evidence against Speciation by Reinforcement." It was not meant to consider the quite distinct question of the occurrence and significance of reinforcement *within* a species. Although I and others[5,6] find the evidence in favor of this idea to be weak, it is a separate issue and was not covered by my title. Another general point which should be kept in mind is that no undisputed case of speciation through reinforcement can be pointed to, despite the model's uncritical endorsement by some authors[7] (p. 576). For this reason I classify it as a Class II model[8] (i.e., a purely hypothetical model). These few points at once provide answers to a number of Littlejohn's critical remarks. For example, on p. 324 he wrote, "Paterson (1978) did not discuss those cases of reproductive character displacement [= reinforcement] that have been analyzed in detail (e.g., Blair, 1955; Littlejohn, 1965; Fouquette, 1975; Ralin, 1977)." Although these examples are all concerned with reinforcement within a species rather than speciation through reinforcement, it is still inaccurate and misleading to imply that I neglected entirely these studies of intraspecific reinforcement in frogs. My reference 32 was to a paper by the late Jasper Loftus-Hills,[5] an experienced specialist on the subject of reinforcement in frogs. His opening words are: "Reproductive character displacement is still considered by many (e.g., Levin, 1969; Cody, 1973) to be a widespread phenomenon in spite of the fact that evidence for its existence is weak; all purported cases can be criticized, usually on a number of grounds (Grant, 1972; Thielke, 1969; Alexander and Loftus-Hills, in prep.). Littlejohn's study (1965) of two species of tree frogs remains the only relatively convincing, documented case among animals." I chose to cite Loftus-Hills' opinion because he had himself published on the pair of species reported on in Littlejohn (1965) and had reviewed the literature on the subject up to 1975, thus including Blair (1955) and Fouquette (1975; see the author's acknowledgment). I did not cover Ralin (1977) because it had not reached England at the time of my talk. Littlejohn's recent review does not consider Ralin's paper nor even cite Loftus-Hills' critical note.

Before proceeding further there is another important general point to discuss. Littlejohn's choice of words is often so careless as to misrepresent my position in fundamental ways. On p. 303, for example, he attributes to me a view which I strongly oppose: "Paterson (1978) stressed this positive nature of mate *selection*, and described such processes and devices as specific mate recognition systems (SMRS)" [my italics]. I have never regarded the function of the specific-mate recognition system as mate *selection*. Mate selection is fundamentally different from specific-mate recognition and carries very different connotations to which I do not subscribe. This can be illustrated by quoting the sentence with which he opens the paragraph under discussion: "One point that emerges from this consideration of mate selection is its positive nature (i.e., the active selection of an appropriate mate) and its extension to notions of sexual selection." This is quite so, and it illustrates most graphically why I draw the sharpest possible distinction between mate selection and specific-mate recognition: I wish to accept no greater a commitment to conscious judgment in sexual organisms than is involved in the recognition of an antigen by an antibody, unless compelled by evidence to do so. All sexual eukaryotes appear to possess an SMRS, relatively simple though it may often be. The SMRS, thus, seems to offer a universal basis for conceiving species within this group. However, my credulity is tested to the full when asked to consider mate selection in very many plants, fungi, protistans, and even animals like oysters. I shall later draw attention to Littlejohn's usage of the term "sexual selection." In the same paragraph he continues, "The correlation between positive assortative mating, which results from the active processes performed by breeding individuals, and sexual or ethological isolation is obvious." Since my ideas have been mentioned in this paragraph, I shall state explicitly the interpretation of this sentence which I am prepared to accept. I accept that positive assortative mating is brought about by the functioning of the SMRS and that this has the fortuitous consequence that populations whose members possess distinct specific-mate recognition systems do not exchange genes. In other words, the function[9] of the SMRS is to facilitate the achievement of syngamy under the conditions which generally prevail in the preferred habitat of the organisms. Any "reproductive isolation" which incidentally results must, thus, be recognized as an effect.[9]

According to my views,[2,4] the SMRS comprises a subclass of the broader class of fertilization mechanisms. Each component of the SMRS involves signaling between potential mating partners (or their cells) and leads to the bringing together of appropriate mating partners (or their cells) and, ultimately, syngamy. The characters of the SMRS are not isolating mechanisms, nor are they mate-selecting mechanisms as Littlejohn believes. With Williams[9] I believe that "mechanism" should be reserved for characters functioning in a role for which they were selected.

It follows from the recognition concept of species that new species arise as incidental consequences of adaptive evolution entailing individual selection.[2,4,8] In contrast, the isolation concept of species (= biological [p. 54] species concept) implies that species are biotic adaptations. Some authors make this quite explicit,[10,11] but with others the commitment is more covert and is revealed by incautious statements relating to the "good of the species," or to isolating mechanisms acting to protect the "integrity of the species." A serious inconsistency in Dobzhansky's theoretical edifice is that he regarded species as "adaptive devices,"[10] which is clearly a group selectionist viewpoint, and yet he consistently[12,13] advocated species formation in accordance with the reinforcement model of speciation, which involves only individual selection.

To conclude this general statement, the essential features of the recognition concept can be stated as follows: The members of a species share a common specific-mate recognition system. The raison d'être of an SMRS is to ensure effective syngamy within a population of organisms occupying their preferred habitat. The characters of the SMRS are adapted to function efficiently in this preferred habitat. Hence, it follows that a new species will have arisen when all members of a small, isolated subpopulation of a parental species have acquired a new SMRS, which facilitates the achievement of syngamy under the new conditions and which, quite fortuitously, makes effective signaling impossible between members of the daughter and parental populations.

Since the characters of the SMRS are adapted to function most efficiently under the typical conditions of the preferred habitat, speciation is the incidental consequence of adaptation to a new environment. Species according to the recognition concept are not relational as in the isolation concept. Species are defined without reference to other species, and no mention is made of the need to preserve their integrity. It should be stressed that the ability of members of a species to remain in their preferred habitat has a powerful homeostatic influence, because it maintains the SMRS and other adaptive characters under stabilizing selection. The outstanding studies of Coope and his colleagues[14] on the movements of beetle populations in response to the fluctuating climatic conditions of the Pleistocene of Europe and America provide very strong support for this view. This expectation from the recognition concept has important implications for Eldredge and Gould's views[15] on the patterns of evolutionary change and on theoretical models such as Van Valen's Red Queen hypothesis[16] (see below).

A MAJOR OBJECTION TO SPECIATION BY REINFORCEMENT

In 1957 Moore[17] published an important paper in which he criticized the model of speciation through reinforcement. Although many of his examples are no longer acceptable, one major criticism still seems to be valid and unanswered. In fact, this difficulty was first noticed by Dobzhansky[12] (p. 318). What exactly was Moore examining in his major criticism? He makes this quite clear at the end of p. 334: "Let us . . . postulate two populations, C and D, which were isolated for some time during which they diverged genetically to such an extent that any hybrids they might form would show reduced viability. Let us further assume that C and D reestablish contact and in the zone of overlap they hybridize. Since we have postulated that hybrids have reduced viability, let us assume with Dobzhansky, that natural selection would promote the development of isolating mechanisms." He then pointed out that "these isolating mechanisms would have selective advantage in the zone of overlap but not elsewhere." In fact, in relation to the unmodified members of the reinforced population, the newly acquired isolating mechanisms will inhibit intrapopulational mating and thus be disadvantageous (see below). How did Littlejohn deal with this problem? While seeming to answer it he actually avoids doing so. On p. 323 he states that the problem

> was considered by J. A. Moore (1957) to be a major criticism of reproductive character displacement because the [isolating] mechanisms appeared to be general characteristics of biological species, whether in sympatric or allopatric populations. The assumption that they are *ad hoc* mechanisms, and hence should be restricted to sympatric zones, presumably is based on the notion that all speciation occurs by way of geographic separation of large portions of a species' range. But it is now plausible to consider other processes of speciation (Bush, 1975). Thus localized sympatric speciation, followed by expansion into allopatric areas, and coupled with the stability associated with sexual selection, could account for the uniformity of mate-selecting systems.

This argument requires careful scrutiny if we are seeking a true evaluation of Moore's objection. First, it should be noted that Moore was not seeking to understand why "isolating mechanisms" are uniformly fixed within species populations. He observed that it was, in fact, empirically so, but wished to understand how this state could be achieved if speciation occurs generally in accord with Dobzhansky's reinforcement model.[12] He could not see how alleles selected in the zone of overlap of C and D could spread to fixation in the allopatric parts of the ranges of C and D. Although it is obvious that this is what Moore wanted explained, Littlejohn never addresses this specific problem. Instead he sets out to explain why species are uniform with respect to their isolating mechanisms. In any case, it is curious that Littlejohn should resort to so dubious a Class II model as sympatric speciation[3,8,18,19] in order to achieve this. What Littlejohn means by "the stability associated with sexual selection" is hard to understand, for it has long been recognized[20] (p. 152) that sexual selection is a positive feedback system and is, thus, unstable. Perhaps he really means "mate recognition," which is stable,[2] rather than "sexual selection." Littlejohn then went on to provide a second explanation of the uniform occurrence of isolating mechanisms throughout the species. This involves a "closed population" model of speciation by reinforcement, which, of course, also fails to address the problem posed by Moore. This model will be returned to in the next section for the perspective it provides on Littlejohn's criticisms. From this discussion and comment it is clear that Littlejohn has not met Moore's objection.

The study of *Litorea* (= *Hyla*) *ewingi* and *L. verreauxi*,[21] which Littlejohn felt that I had neglected, can actually be used to provide an example to illustrate Moore's main critical point. These frogs have been studied in allopatry and sympatry,[21-23] and these data constitute the only relatively convincing case of reproductive character displacement among animals, according to Loftus-Hills.[5] What Moore would want to know is how the alleles which are selected to reinforce in the zone of overlap ("western sympatry") can spread to fixation within *verreauxi* as a whole. Littlejohn and Loftus-Hills[22] actually provide evidence which demonstrates very graphically the difficulty in accounting for the spread of reinforcing alleles into the allopatric parts of *verreauxi*'s range. They demonstrated that *L. verreauxi* from the area of "western sympatry" have calls different from those in "remote allopatry," and that these different calls are "highly effective as a potential isolating mechanism." In other words, the alleles which were selected to reinforce call differences in the area of sympatry with *L. ewingi* are disadvantageous in the allopatric parts of *verreauxi*'s range. Indeed, as Loftus-Hills[5] has stated (see above), there are a number of aspects of this study which can be criticized. As things stand, this study of Littlejohn's may, perhaps, provide some evidence for intraspecific reinforcement, but has little force in supporting a case for speciation through reinforcement in the face of Moore's objection.

REINFORCEMENT ACCORDING TO THE
CLOSED POPULATION MODEL

It should now be clear why the evidence for speciation by reinforcement at a parapatric zone, narrow or broad, is not compelling. This leaves only the "closed population" model of speciation by reinforcement for consideration. According to this model, a small daughter population of a species becomes isolated in a habitat different from the preferred habitat of the members of the parent species, and hence the members are subject to directional selection. [p. 55] Following the divergence of its members, the daughter population, for example, becomes surrounded by the expanding parental species. It is at this stage that reinforcement occurs. The conditions of this scenario do not seem to be improbable. They might follow from extreme fluctuations in climate such as occurred during the Pleistocene. In fact, it is such a model that Littlejohn invoked on p. 323 of his paper when attempting to deal with Moore's difficulty (see above): "Similarly, speciation in a small peripheral isolate, followed by a phase of complete sympatry with the parental species during which reinforcement occurred, and subsequently expansion of range of the new species into allopatric areas, again under the influence of stabilizing sexual selection, could explain the uniformity of homogamic mechanisms." (His usage of sexual selection has already been criticized.) It is, therefore, extraordinary that Littlejohn should then have immediately gone on to criticize my[2] invoking of this model for discussion. Much of what he writes in criticism is irrelevant: I am criticized for not specifying the area occupied by the daughter population. If this is vital, why did he not specify it either? I was not justified in attributing the basis of the hybrid sterility to a translocation which had become fixed in the daughter population; Littlejohn did not specify any cause. In fact, the exact basis of the sterility is not at all critical. In a comparable model,[26] which in 1969 met with Littlejohn's approval, Wilson specified a two-allele basis for the disadvantage. I specified the translocation basis for a number of reasons. First, it met the conditions specified for the model[7] I quoted; secondly, it facilitated picturing what was occurring; thirdly, it enabled me to demonstrate the point that evolutionists invoke reinforcement when considering an interpopulational scenario, but they invoke the algebraic model for heterozygote disadvantage when an intraspecific scene is drawn, despite the fact that the basis is exactly the same. This was meant to demonstrate in a devious way the strong subjective bias in all such discussions, regardless of whether the author is Fisher, Dobzhansky, or Littlejohn. This same subjectivity was again referred to at the end of my paper, when I inserted a quotation from Crosby,[27] who referred to it as an idealistic view of species.

My algebraic study revealed that natural selection would act to eliminate the *cause* of the heterozygote disadvantage. This is a much more likely course than the reinforcement alternative, because it does not require the ready availability of appropriate reinforcing alleles in the small daughter population, nor does it encounter difficulties due to the coadaptation of the male and female components of the mate recognition system.[2] It simply involves selection against the allele or alleles, translocation, pericentric inversion, etc., which determines the infertility or partial infertility of the hybrids. Though Littlejohn asserts limitations for my quantitative model, he does appear to accept its findings within these limitations, for he wrote, "It seems, however, that his [Paterson's] conclusion can only apply to the closed sympatric model." However, Littlejohn's reluctance to apply the findings is clear, for he does not even consider the outcome my model predicts when he advocates his own closed population model on p. 323. Why not, one wonders?

It is misleading that Littlejohn does not inform the readers of his review that I had provided some empirical support for my conclusions (refs. 18 and 19)[2] and that I countered a number of arguments which have been used to support the idea of speciation through reinforcement. Of these I shall mention only the *Drosophila* and *Zea* experiments which ostensibly support experimentally the feasibility of the reinforcement model. In an earlier review of the systematic significance of isolating mechanisms, Littlejohn[28] wrote, "That the process [reinforcement] certainly occurs has been amply demonstrated in experimental populations of *Drosophila*; for example, Koopman (1950) and Kessler (1966), through a program of direct selection, were able to strengthen ethological isolation between *D. pseudoobscura* and *D. persimilis*. Wallace (1954) and Knight et al. (1956) demonstrated that homogamic mating preference could be increased in mutant strains of *D. melanogaster*." In my 1978 [chap. 3] paper[2] I pointed out that all such experiments which I had examined were flawed in their design. In all cases the experiments had maintained parity in the numbers of the two experimental populations over the generations. This prevented the elimination of the cause of the hybrid disadvantage; thus, it is not surprising that some reinforcement occurred. Even so, it was not great, and no evidence was ever produced to show that the new state was stabilized.

I also drew attention to the same flaw in Crosby's[27] (p. 259) computer simulation of reinforcement: "Within a few generations, one sub-species had eliminated the other. Although the population size had been kept constant, no provision had been made for maintaining equality of A and B. Inevitably, chance led to inequality and provided another example of a highly metastable situation with disequilibrium having very strong positive feedback characteristics . . . Here, it was merely a nuisance to be eliminated, and this was done by the occasional addition of 16G or 16g plants." It is thus clear that when Crosby eliminated the nuisance he also blocked the alternative and more probable course open to natural selection. Littlejohn does not consider any of these points in his latest review except to write: "Because . . . there is a potential for migration in most natural systems, the laboratory studies, and Crosby's later models . . . , it is difficult to see the general relevance of Paterson's conclusion." First, one notes that no provision for migration was made in his own model for a closed population (p. 323), and yet he believes that reinforcement will be its outcome. Secondly, it is very misleading indeed to invoke migration in the way Littlejohn does. The adjusting of numbers, as was done both by the experimenters and by Crosby, cannot glibly be said to be equivalent to migration, for it has no counterpart in nature as far as I know. If it occurred in nature it would involve a form of differential migration which would ensure that neither population declines to extinction when the coefficient of selection, s, is unity ($s = 1$). It is evident that there is a need for close attention to detail if such inadequacies are to be detected in Littlejohn's review.

THE SMRS IN RELATION TO THE PREFERRED HABITAT

A statement by Littlejohn (p. 324) requires comment because it undermines a fundamental component of the recognition concept of species, the basic role of habitat preference as a species-specific character. He wrote: "Nor did he [Paterson] consider the influence of differing reproductive environments on the allopatric divergence of mate-selecting mechanisms." As already emphasized, I did not consider "mate selecting mechanisms" at all. Secondly, it should be made clear that "reproductive environment" is not simply the environment in which reproduction occurs, but, in frogs, refers particularly to the sound environment in which auditory communication between the sexes must occur. Littlejohn[28] accepts that even noncongeners can interfere with auditory communication. This interference, he thinks, may provide the selective basis which will result in the divergence of what he calls "mate-selecting mechanisms."

What I did consider was the influence of different environments on the allopatric divergence of specific-mate recognition systems[2] (p. 370): "These properties enable an SMRS to be modified by selection to fit a new environment, for example." Later[4] I provided an even clearer statement on the crucial role which I accept for the preferred habitat of a population in shaping the SMRS, which, in turn, is the basis of speciation. I do not accept that the calls of allogeneric species of frogs constitute a selective basis for speciation. The particular constellation of species at a given breeding site will vary within one night, within one season, and from season to season. As a formula for speciation this seems to be quite improbable. To the best of our current knowledge speciation at its fastest would [p. 56] take several hundred generations. Is a local population of a frog species, which is suffering from auditory interference from another species, likely to survive as long as this before it has adjusted to the interfering call? Or will it simply die out locally or move to another breeding site which is free from interference? In semitropical areas twelve or more species may share a breeding station without evident indications of interference (N. I. Passmore, personal communication). However, this hypothesis of Littlejohn's is surely open to testing and falsification.

In contrast, the association of the Cape Ghost frog (*Heleophryne purcelli* Sclater)[29] with fast-running mountain streams and their attendant high noise levels is quite likely to have led to the selective shaping of the call for effective communication between the sexes under these adverse conditions. The selective pressure is very stable and poses no theoretical problem.

Littlejohn pointed out that sufficient variation exists in anuran calls for natural selection to be effective. Read in context this seems to suggest that I thought otherwise. In fact, I clearly accepted this point on p. 370,[2] and, furthermore, I pointed out that signaling systems (SMRS) were subject to stabilizing selection because of the coadaptation which exists between the signaling and receiving systems of the two sexes.

These further comments, those to be found above under the heading "General Considerations," and the statements made in my earlier papers[2,4] should make plain the fundamental role I assign to the habitat in speciation and the stabilization of species. The preferred habitat of a species approximates to the habitat in which the species arose. Habitat preference is, thus, a fundamental, species-specific character, because restriction to the preferred habitat keeps all the species-specific adaptive characters of the members of the species under stabilizing selection. These adaptive characters include the components of the SMRS. This provides a robust explanation for the stasis which paleontologists[15,30] observe to follow speciation events (i.e., the equilibrium of "punctuated equilibrium"). As already mentioned, the work on Coleoptera[14] at particular sites during the Pleistocene of Europe provides graphic support for these conclusions. It is, perhaps, appropriate to emphasize here that the case for the "punctuated equilibrium"[15,30] pattern of variation in the fossil record is particularly well supported when the fossils carry characters of importance in the SMRS of the species. Such characters enable speciation events to be detected with great reliability. Examples are the external genitalia in Coope's beetles, and the horns in Vrba's antelope.[30,31]

The relationship between members of a species and their preferred habitat is also of relevance to Van Valen's Red Queen hypothesis,[16] as already stated. According to Van Valen, natural selection acts mainly "to enable the organisms to maintain their state of adaptation rather than to improve it" in the face of constant environmental decay. Both Coope's studies and the theory relating organisms to their preferred habitat would suggest that organisms would be more likely to move to better areas of the preferred habitat than remain immobile as their environment changes, unless they were prevented from doing so. One wonders whether the Red Queen hypothesis applies only to populations which cannot retreat through their path of retreat being blocked by some natural barrier such as a mountain range. Coope commented on the absence of speciation among the beetles which he studied and did suggest that speciation was likely to occur only when populations were blocked from retreating.

DISCUSSION

Evolutionary biology is a complex field calling for skill, expertise, and experience in many fields, not least in philosophy and writing. Any case under criticism needs to be presented fairly and in its strongest form if science is to be advanced. Awkward facts must be presented in reviews as well as convenient ones. Hypotheses need to be tested in such a way that they are put at real risk. Ad hoc shoring up of hypotheses should be avoided. Words and terms should not be degraded as, for example, has occurred with terms such as "niche" and "choice" and concepts such as "sexual selection." The rules of scholarship require one to use the original literature rather than secondary sources; references provided by authors need to be consulted and checked to see that they say what they purport to say. All these rules should be followed by evolutionary biologists, but they are not. This became evident, for example, when a notable legal mind became interested in evolutionary theory, examined its writings, and revealed what he found for the benefit of others.[32] Today when evolutionary theory is again under attack, it is vital that we avoid the slipshod evolutionary writing and logic which has been so widely practiced in the last fifty years.[33]

I have examined in depth Littlejohn's criticisms of aspects of a paper of mine and of aspects of a paper by J. A. Moore. I found that he has misrepresented my case and has not even addressed himself to the problem posed by Moore. Moore's objection has been very generally left unanswered for more than thirty years because it undermines an hypothesis of importance to those who support the isolation concept of species (e.g., Dobzhansky[13]). If it is ever to be answered and thus disposed of, let it be done proficiently.

Littlejohn agrees with Williams,[9] Ghiselin,[34] and Paterson[2,8,35] about the shortcomings of the current species concept, the isolation concept.[4] However, his solution to these difficulties raises many difficulties of its own. He invokes, in place of my specific-mate recognition system and fertilization mechanisms, what he calls homogamic systems which are very diverse in basis, some acting to bring about positive assortative mating purely incidentally ("effects"), but others doing so as adaptations, though he evidently believes this is achieved through *mate selection* in contrast to my view that nothing more than *mate recognition* is involved. Thus, Littlejohn's homogamic mechanisms are as diverse as Dobzhansky's isolating mechanisms. As a consequence of this, he (p. 328) finds it impossible to bring them under a single convenient and concise term, which is not surprising considering their heterogeneity. This is in sharp contrast to the recognition concept of species. According to this concept, members of a species mate positively assortatively through the functioning of their shared specific-mate recognition system. The SMRS of a species is a subset of the broader set, the fertilization mechanisms of the species, and it comprises a coadapted signal-response reaction chain involving the mating partners. The raison d'être of an SMRS is to ensure effective syngamy within a population of a species in which the individual members are occupying their preferred habitat. Although it is obvious that a number of extraneous factors can influence fertilization adventitiously as "effects," they must obviously be excluded. Mayr[3] (p. 91), speaking of isolating mechanisms, said that if we are to discuss intelligently their nature and origin we must exclude extraneous elements, and so it must be with fertilization mechanisms and specific-mate recognition systems.

In this article I have devoted very much more space to answering

Littlejohn's objections to my paper[2] than he took to make them. This apparent prodigality requires some explanation. In fact, besides my response to the criticisms of my ideas and Moore's, I have extended my writings on the recognition concept of species and provided some indication of their significance for evolutionary biology. This means that a more comprehensive statement of the concept will now be available if this paper is read together with my earlier ones[2,4,8] and a recent one by Passmore.[29] I have also been able to substantiate further my earlier[8] stated belief that we need a more consistent logic if we are to reduce dispute and disagreement and improve understanding and comprehension in evolutionary biology. This must, in any case, be done in order to avoid aiding unnecessarily the current obscurantist antievolutionary campaign in progress overseas, which relentlessly exploits dispute and disagreement in evolutionary biology.[33] [p. 57]

Finally, I hope that my analysis of parts of Littlejohn's critical review will encourage a more extended critical reading of the work as a whole.

I owe thanks to Marc Centner, Anthony Gordon, Christopher Green, David Lambert, Judith Masters, Neville Passmore, Hugh Robertson, and Elisabeth Vrba for their criticism and comments.

REFERENCES

1. Littlejohn, M. J. 1981. Reproductive Isolation: A Critical Review. Chap. 15 in *Evolution and Speciation*, ed. W. R. Atchley and D. S. Woodruff, 298–334. Cambridge: Cambridge University Press.
2. Paterson, H.E.H. 1978. More evidence against speciation by reinforcement. *South African Journal of Science* 74:369–71. [This volume, chap. 3.]
3. Mayr, E. 1963. *Animal Species and Evolution*. Cambridge, Mass.: Belknap Press of Harvard University Press.
4. Paterson, H.E.H. 1980. A comment on "Mate Recognition Systems." *Evolution* 34:330–31. [This volume, chap. 4.]
5. Loftus-Hills, J. J. 1975. The evidence for reproductive character displacement between the toads *Bufo americanus* and *B. woodhousii fowleri*. *Evolution* 29:368–69.
6. Thielke, G. 1969. Geographic Variation in Bird Vocalizations: Their Relations to Current Problems in Biology and Psychology. Essays presented to W. H. Thorpe. Chap. 14 in *Bird Vocalizations*, ed. R. A. Hinde, 311–38. Cambridge: Cambridge University Press.
7. Ayala, F. J., M. L. Tracey, D. Hedgecock, and R. C. Richmond. 1974. Genetic differentiation during the speciation process in *Drosophila*. *Evolution* 28:576–92.
8. Paterson, H.E.H. 1981. The continuing search for the unknown and unknowable: A critique of contemporary ideas on speciation. *South African Journal of Science* 77:113–19. [This volume, chap. 5.]

9. Williams, G. C. 1966. *Adaptation and Natural Selection*. Princeton, N.J.: Princeton University Press.

10. Dobzhansky, T. 1976. Organismic and Molecular Aspects of Species Formation. In *Molecular Evolution*, ed. F. J. Ayala, 95–105. Sunderland, Mass.: Sinauer Associates.

11. White, M.J.D. 1978. *Modes of Speciation*. San Francisco: W. H. Freeman.

12. Dobzhansky, T. 1940. Speciation as a stage in evolutionary divergence. *American Naturalist* 74:312–21.

13. Dobzhansky, T. 1970. *Genetics of the Evolutionary Process*. New York: Columbia University Press.

14. Coope, G. R. 1979. Late Cenozoic fossil Coleoptera: Evolution, biogeography, and ecology. *Annual Review of Ecology and Systematics* 10:247–67.

15. Eldredge, N., and S. J. Gould. 1972. Punctuated Equilibria: An Alternative to Phyletic Gradualism. Chap. 5 in *Models in Paleobiology*, ed. T.J.M. Schopf, 82–115. San Francisco: Freeman, Cooper and Company.

16. Van Valen, L. 1973. A new evolutionary law. *Evolutionary Theory* 1:1–30.

17. Moore, J. A. 1957. An Embryologist's View of the Species Concept. In *The Species Problem*, ed. E. Mayr, 325–38. Publication No. 50. Washington, D.C.: American Association for the Advancement of Science.

18. Futuyma, D. J., and G. C. Mayer. 1980. Non-allopatric speciation in animals. *Systematic Zoology* 29:254–71.

19. Jaenike, J. 1981. Criteria for ascertaining the existence of host races. *American Naturalist* 17:830–34.

20. Fisher, R. A. 1958. *The Genetical Theory of Natural Selection*. 2d ed. New York: Dover Publications.

21. Littlejohn, M. J. 1965. Premating isolation in the *Hyla ewingi* complex (Anura: Hylidae). *Evolution* 19:234–43.

22. Littlejohn, M. J., and J. J. Loftus-Hills. 1968. An experimental evaluation of premating isolation in the *Hyla ewingi* complex (Anura: Hylidae). *Evolution* 22:659–63.

23. Watson, G. F., and A. A. Martin. 1968. Postmating isolation in the *Hyla ewingi* complex (Anura: Hylidae). *Evolution* 22:664–66.

24. [Not used.]

25. [Not used.]

26. Wilson, E. O. 1965. The Challenge from Related Species. Chap. 2 in *The Genetics of Colonizing Species*, ed. H. G. Baker and G. L. Stebbins, 7–27. New York: Academic Press.

27. Crosby, J. L. 1970. The evolution of genetic discontinuity: Computer models of the selection of barriers to interbreeding between species. *Heredity* 25:253–97.

28. Littlejohn, M. J. 1969. The Systematic Significance of Isolating Mechanisms. In *Systematic Biology*, ed. C. G. Sibley, 459–82. Washington, D.C.: National Academy of Sciences.

29. Passmore, N. I. 1981. The relevance of the specific mate recognition concept to anuran reproductive biology. *Monitore zoologico italiano (N.S.) Supplemento* 15:93–108.

30. Vrba, E. S. 1980. Evolution, species and fossils: How does life evolve? *South African Journal of Science* 76:61–84.
31. Vrba, E. S. 1984. Evolutionary Pattern and Process in the Sister-Group Alcelaphini-Aepycerotini (Mammalia: Bovidae). In *Living Fossils*, ed. N. Eldredge and S. M. Stanley, 62–79. New York: Springer-Verlag.
32. Macbeth, N. 1971. *Darwin Retried*. New York: Dell.
33. Flew, A. 1981. Fight the obscure. *Nature* 292:192.
34. Ghiselin, M. T. 1974. *The Economy of Nature and the Evolution of Sex.* Berkeley: University of California Press.
35. Paterson, H.E.H. 1973. Animal species studies. Pp. 31–36 in Animal and plant speciation studies in Western Australia, by H.E.H. Paterson and S. H. James. *Journal of the Royal Society of Western Australia* 56:31–43. [See this volume, Author's Preface.]

9 Darwin and the Origin of Species

Paterson, H.E.H. 1982. Darwin and the origin of species.
South African Journal of Science 78:272–75.

*In 1982 biologists across the world remembered the death of
Charles Darwin. The University of the Witwatersrand
offered a Darwin Evening to the public of Johannesburg.
The speakers were Phillip Tobias, Elisabeth Vrba, and I, in
full evening dress. A significant feature of the evening was
that our chairman was the late Albert Geiser, whose chair at
Wits had been created some years earlier when he had been
dismissed by the University of Pretoria from his chair of
divinity after having been found guilty of heresy by his
church. Phillip Tobias and I felt that it would be appropriate
in 1982, the University of the Witwatersrand's sixtieth
anniversary year, that Albert should chair this evening of
homage to Darwin. It certainly said something about Wits.*

*This paper enabled me to explore possible subliminal
influences which seemed to make the isolation concept very
appealing to many, despite its patent disadvantages (see also
Paterson, 1988, p. 65 [chap. 14]). Some of my points were
probably of special interest to Albert.*

To discuss Charles Darwin's views on species and how they originate
provides me with a suitable theme to celebrate this university's sixtieth
year of life. This is because I can use Darwin's main ideas to illustrate
any true university's major concerns. These are the search for truth,
wherever it leads; the pursuit of ideas without regard to their end;
and education.

Education is specially relevant to my theme. Education, when fully
achieved, is tantamount to the freeing of the mind and the achievement of human awakening. The freeing of the mind has a shattering
effect. Darwin's mind was probably not a phenomenal one, but it was
a relatively free one, free of many of the cultural constraints that
limited the often more brilliant minds of his contemporaries. His

revolutionary ideas resulted from his freedom to think the unthinkable, rather than the greatness of his intellect. Man reaches full humanity when he acquires the ability to think independently, for thinking, especially abstract thinking, is *the* human attribute, as was long ago recognized when our species was named *Homo sapiens*.

All these ideas are well illustrated by the work of Darwin, for he probed some of the great problems of the world, including that of man's own nature. Concern for such thoughts tests the real humanity of the investigator, for it demands intellectual honesty of the highest order. I believe that few have endured this trial as well as Darwin. Anyone who doubts the power of ideas, or who tends to minimize the roughness of the path of truth, will be instructed by following Darwin's story.

Darwin's great book, *On the Origin of Species*, was published when he was fifty and already well known, his status having been assured by his election as a Fellow of the Royal Society of London. With little advertising the first printing sold out on the day of publication, 24 November 1859 [Fig. 1].

Why did a sober, unemotional work of scholarship elicit such a response from the inhabitants of Victorian London? Almost certainly because of its provocative title, which proclaimed that the author accepted that species had not been specially created by God in accordance with one or other of the accounts of the Creation to be found in the book of Genesis. Londoners seem to have been eager to establish what these views implied for their own species.

THE BACKGROUND TO THE *ORIGIN*

Let us try to establish what most people thought about the origin of species in 1859. William Paley, in 1802, had written a most persuasive and influential book, *Natural Theology*, which Darwin once boasted he had known almost by heart. Let us notice how Paley argued for the existence of a Grand Watchmaker from observing the perfection of the human eye:

> Were there no example in the world of a contrivance, except that of the eye, it would alone be sufficient to support the conclusion, which we draw from it, as to the necessity of an intelligent Creator.

He then listed the various remarkable characteristics of the eye and ended as follows:

> These provisions compose altogether an apparatus, a system of parts, a preparation of means, so manifest in their design, so exquisite in their contrivance, so successful in their issue, so precious, and so infinitely

ALBEMARLE-STREET,
Nov. 1859.

THIS DAY.

ON THE ORIGIN OF SPECIES
by Means of Natural Selection ; or, the Preservation of Favoured
Races in the Struggle for Life. By CHARLES DARWIN, M.A.,
Author of ' Naturalist's Voyage Round the World.' Post 8vo. 14s.

THOUGHTS on GOVERNMENT
and LEGISLATION. By LORD WROTTESLEY, F.R.S.
Post 8vo. 7s. 6d.

HISTORICAL EVIDENCES of the
TRUTH of the SCRIPTURE RECORDS STATED ANEW,
with Special Reference to the Doubts and Discoveries of Modern
Times. The Bampton Lectures for 1859. By REV. GEORGE
RAWLINSON, M.A. 8vo. 14s.

The ARCHÆOLOGY of BERK-
SHIRE : an Address delivered at Newbury, Sept. 1859. By the
EARL OF CARNARVON. Post 8vo. 1s.

MODERN SYSTEMS of FORTIFI-
CATION, examined with Reference to the NAVAL, LITTORAL,
and INTERNAL DEFENCE of ENGLAND. By GENERAL
SIR HOWARD DOUGLAS, Bart. Plans. 8vo. 12s.

SCIENCE in THEOLOGY. Sermons
Preached before the University of Oxford. By REV. ADAM S.
FARRAR, Fellow of Queen's College. 8vo. 9s.

LORD BYRON'S COMPLETE
WORKS, with Notes and Illustrations by JEFFREY, HEBER,
WILSON, MOORE, GIFFORD, LOCKHART, &c. With Por-
trait and Engravings. Royal 8vo. 9s. ; or cloth. 10s. 6d.

SELF-HELP. With Illustrations of
Character and Conduct. By SAMUEL SMILES, Author of
' Life of George Stephenson.' Post 8vo. 6s.

A MANUAL of the ENGLISH CON-
STITUTION : a Review of its Rise, Growth, and Present State.
By DAVID ROWLAND. Post 8vo. 10s. 6d.

Figure 1. John Murray's advertisement of the first edition of the *Origin of Species*, issued on November 24, 1859, at Albemarle Street.

beneficial in their use, as, in my opinion, to bear down all doubt that can be raised upon the subject. . . . If there were but one watch in the world, it would not be less certain that it had a maker.

This kind of argument was in those days all-prevalent and is still encountered today.

In 1857, just before the *Origin* was published, J. D. Dana, professor of natural history and geology at Yale University, wrote,[1] "Species are units fixed in the plan of creation . . . divine appointments which cannot be obliterated."

Professor J. W. Dawson,[2] of McGill College, Montreal, in his turn, thought that "the species . . . with all its powers and capacities for reproduction, is that which the Creator made; His unit in the work, as well as ours in the study."

In the course of reviewing the *Origin*, Wollaston, the British naturalist, stated that[3] "the opinion among naturalists that species were independently created, and have not been transmuted one from the other, has been hitherto so general that we might call it an axiom."

The Creator was interpreted as having made provision to protect his handiwork, the species, from degradation by hybridization. In a popular journal of the time can be found:[4] "God forbids hybrids to breed"; and in a religious review:[5] "the sterility of true hybrids afford another evidence of the jealousy with which the Creator regards all attempts to introduce confusion into His perfect plan." The same point was made by Darwin in introducing his chapter 8 on "Hybridism" in the *Origin of Species*: "The view generally entertained by naturalists is that species, when intercrossed, have been specially endowed with the quality of sterility, in order to prevent the confusion of all organic forms" (i.e., by hybridization). Darwin, himself, of course, did not accept this view and at once set about discrediting it.

Of special significance here are the views of Edward Blyth. Darwin especially esteemed the opinions of Blyth, as can be seen from this testimonial from the *Origin* (1859:18): "Mr Blyth, whose opinion, from his large and varied stores of knowledge, I should value more than that of almost any one." In 1835 Blyth wrote[6] the following passage, which demonstrates a clear appreciation of the principle of natural selection in the form of stabilizing selection, but also proves that he regarded the role of selection as the preservation of the specially endowed properties of species:

> In like manner, among animals which procure their food by means of their agility, strength, or delicacy of sense, the one best organized must always obtain the greatest quantity; and must, therefore, become physically the strongest, and be thus enabled, by routing its opponents, to transmit its superior qualities to a greater number of offspring. The same law, therefore, which was intended by Providence to keep up the typical qualities of a species, can easily be converted by man into a means of raising different varieties; but it is also clear that, if man did not keep up these breeds by regulating the sexual intercourse, they would all naturally soon revert to the original type.

Along similar lines, a correspondent to *Good Words* in 1862 wrote:

> When God saw that every living creature which he made was good, we cannot doubt that the type of each of them was perfect. The struggle for life, therefore, is to *prevent*, and not to promote a change in the original form.

Such views can be traced back at least to John Ray (1627–1705). Linnaeus, in his *Critica Botanica*, wrote:

> All species reckon the origin of their stock in the first instance from the veritable hand of the Almighty Creator: for the Author of Nature, when

He created species, imposed on His creations an eternal law of reproduction and multiplication within the limits of their proper kinds. He did
indeed in many instances allow them the power of sporting in their out-
ward appearance, but never that of passing from one species to another. . . . And so I distinguish the species of the Almighty Creator which
are true from the abnormal varieties of the Gardener; the former I reckon
of the highest importance because of their author, the latter I reject
because of their authors. The former persist and have persisted from the
beginning of the world, the later, being monstrosities, can boast of but a
brief life. [p. 273]

Ray (1727) wrote[7] as follows in *The Wisdom of God Manifested in the
Works of the Creation*, "First of all, because it is the great design of
Providence to maintain and continue every species, I shall take Notice
of the great Care and abundant Provision that is made for securing
this end."

In his *Historia Plantarum* (1686) he had earlier written: "Thus, no
matter what variations occur in the individual or the species, if they
spring from the seed of one and the same plant, they are accidental
variations and not such as to distinguish a species . . . Animals likewise
that differ specifically preserve their distinct species permanently; one
species never springs from the seed of another nor *vice versa*."

All these views can be derived proximately from the Scholastic
culture of Western Europe, and they shared with it the same uneasy
relationship between reason and faith. Something of the Great Design
of Providence mentioned by Ray can be gathered from quotations
from the book of Genesis:

Let us make man in our image, after our likeness:[8] and let them have
dominion over the fish of the sea, and over the fowl of the air, and over
the cattle, and over all the earth, and over every creeping thing that
creepeth upon the earth.

And from a little further on:

Behold, I have given you every herb bearing seed, which is on the face
of all the earth, and every tree, in which is the fruit of a tree yielding
seed; to you it shall be for meat. (See also Genesis 9:1–3.)

In accordance with this view man is clearly something very special,
having been made in God's image and having been granted dominion
over all other organisms. It is, therefore, hardly surprising that many
are reluctant to be degraded from such senior status.

Philosophically, this view of species is idealistic and teleological.
Each species has been assigned a role in the Divine Plan and provided
with characteristics to fit it for that role. These, in turn, are protected
from degradation due either to "sporting" or to hybridization. Each

species, according to the prevailing opinion of 1859, possesses a *purpose* (its role in the Great Design of Providence) and an *integrity* (its properties assigned by the Creator) which are protected by such devices as sterility and by the elimination by natural selection of variants. Such views are compatible with those of Aristotle, representing the European stem of Scholasticism.

This, then, was the background against which the *Origin of Species* appeared in 1859. This background must be understood and appreciated because, as Skolimowski has pointed out,[9] "the difficulties of present biology are more conceptual than empirical, more rooted in our *Weltanschauung* than the actual empirical problems we investigate."

DARWIN'S VIEW OF SPECIES

What view of species was disclosed to the first readers of the *Origin?*

> I look at the term species, as one arbitrarily given for the sake of convenience to a set of individuals closely resembling each other, and that it does not essentially differ from the term variety, which is given to less distinct and more fluctuating forms. (P. 52)

On the same page it is made clear that varieties and species both arise, in Darwin's opinion, through the action of natural selection:

> And I attribute the passage of a variety, from a state in which it differs very slightly from its parent to one in which it differs more, to the action of natural selection in accumulating . . . differences of structure in certain definite directions. Hence I believe a well marked variety may be justly called an incipient species.

How should Darwin's conception of species be regarded in philosophical terms?

Darwin evidently believed that species arose as the incidental consequence of adaptation to a new environment to which the founders had become restricted. He clearly judged species status in taxonomic terms, for he mentions "accumulating differences of structure" in the second quotation above when discussing speciation and says on page 47, "In determining whether a form should be ranked as a species or variety, the opinion of naturalists having sound judgment and wide experience seems the only guide to follow."

The important point, however, is that he certainly did not subscribe to the prevailing teleological view revealed above. On page 245 he contradicts the prevailing view that the quality of sterility has the function of preventing the confusion of organic forms of hybridization:

On the theory of natural selection the case is especially important, inasmuch as the sterility of hybrids could not possibly be of any advantage to them and therefore could not have been acquired by the continued preservation of successive profitable degrees of sterility.

Sterility is not a device to preserve the integrity of a species. It, too, arose as an incidental consequence of adaptation to a new environment. Since Darwin's species lacked a "purpose" and, hence, an "integrity," the need for devices to guard against "confusion" is not apparent. These species are thus conceived in empirical terms, quite devoid of teleology.

Few evolutionists have appreciated how fundamentally different was Darwin's viewpoint on species. However, the point was noticed by philosophers of science. Thomas S. Kuhn wrote in *The Structure of Scientific Revolutions*: "For many men the abolition of that teleological kind of evolution was the most significant and least palatable of Darwin's suggestions. The *Origin of Species* recognized no goal set either by God or nature." And David Hull[10] has pointed out that "Darwin dismissed theological explanations of the origin of species because they were not properly scientific. He also dismissed explanations in terms of 'plans,' divine or otherwise. They, too, were empty verbiage."

Among contemporary evolutionary theorists, George Williams[11] is virtually alone in appreciating that "the species is . . . a key taxonomic and evolutionary concept but has no special significance for the study of adaptation. It is not an adapted unit and there are no mechanisms that function for the survival of the species" (p. 252).

Thus, Darwin's views on species provided a new way ahead, free of the supernatural. We shall see, however, that it was ignored.

WALLACE'S VIEW OF SPECIES

While Darwin espoused a materialistic and empirical approach to species and their origin, Wallace was more theistically inclined and, latterly, much influenced by spiritualism. This commitment to the supernatural was in sharp contrast to Darwin. The following revealing passage is from *Darwinism* (p. 477): [p. 274]

> To us, the whole purpose, the only *raison d'être* of the world—with all its complexities of physical structure, with its grand geological progress, the slow evolution of the vegetable and animal kingdoms, and the ultimate appearance of man—was the development of the human spirit in association with the human body.

Wallace, nevertheless, accepted no commitment to the special creation of man, as is quite evident from the following quotation from the same book (p. 455): "We are compelled to reject the idea of 'special creation' for man, as being entirely unsupported by facts as well as in the highest degree improbable."

Thus Wallace's view of species (man, at least) was clearly teleological. For many years he argued with Darwin over the possibility of sterility in crosses between species arising under natural selection. Wallace, like most naturalists of the period before the publication of the *Origin*, including Blyth, regarded sterility as an adaptation evolved to preserve the specific integrity of species. On the other hand, Darwin interpreted sterility as having arisen as an incidental by-product of two populations diverging under different selective regimes.

THE CURRENT PARADIGM

What is the orthodox view of species today, and from where does it stem? What is its phylogeny?

The concept of species which prevails today achieved general acceptance through the writings of Theodosius Dobzhansky, particularly his book *Genetics and the Origin of Species*, which first appeared in 1937. His view of species has been modified to some degree by Ernst Mayr, but it still remains virtually the same after forty-five years. This is how Dobzhansky saw species: "Unlimited interbreeding of distinct species would result in the submergence of existing genetic systems in a mass of recombinations." (Notice how this view resembles that of the authors, some eighty years earlier, who were concerned about the "confusion of all organic forms" due to hybridization.) In 1951[12] Dobzhansky defined species in genetical terms as follows: Species are "groups of populations the gene exchange between which is limited or prevented in nature by one, or a combination of several, reproductive isolating mechanisms."

Among the isolating mechanisms listed by Dobzhansky we find various forms of sterility. For the greater part of his long career Dobzhansky vacillated over whether sterility could arise under natural selection. At heart, he appeared to be on Wallace's side, but seemed unable to counter Darwin's rudimentary but correct genetic logic given at the start of chapter 8 of the *Origin*. Darwin[13] once wrote in extreme frustration to Huxley, who, like Wallace, was a little slow in appreciating the genetic problems involved: "Nature never made species mutually sterile by selection, nor will men."

That Dobzhansky saw species fulfilling a purpose is evident from the following passage,[14] which reiterates a view which he often expressed:

In general, the working hypothesis which seems to me fruitful is that species are not accidents but adaptive devices through which the living world has deployed itself to master a progressively greater range of environments and ways of living.

Replacing "the living world" by "God," one can almost read off His Divine Plan. It appears that Dobzhansky resembled Wallace by espousing a teleological and idealistic philosophical view of species. In fact, both authors' view owes much more to Blyth than to Darwin.

I confess to experiencing a feeling of shock when I realized that Darwin's great achievement of freeing biology of teleology and the supernatural had been surreptitiously supplanted by the old theological concept deceptively clad in new clothes. Skolimowski[9] neatly summarized the situation as follows: "Though God is denounced and declared non-existent, his shadow still haunts the biologist turned natural philosopher." It is as if the great conceptual revolution had never occurred!

THE FUTURE

For some eighteen years I accepted and practiced the isolation concept of species, but eventually its inconsistencies became too intrusive to accept. How can a concept be espoused when it so clearly is teleological and derived from Scholastic roots? Dobzhansky's view is in fact unsatisfactory in many ways,[15–19] and this has led me to derive an alternative concept which is free from all these objections. I call this the "recognition concept," and Dobzhansky's the "isolation concept."[17]

It is characteristic of members of a species that they generally mate with one another rather than with members of other species. In other words, they mate positively assortatively. This is because every biparental species must possess mechanisms for bringing mates together and achieving copulation. Fertilization is unlikely to be achieved fortuitously! There are many fertilization mechanisms: the genitalia, the form of the sperm, the means by which the coordination of gamete-ripening occurs, and so on. Always there is a class of fertilization mechanisms which entails signaling between the mates, or between their cells. This system I have called the "specific-mate recognition system" (SMRS).[16] Every species possesses its own distinct SMRS, regardless of the kingdom to which it belongs. Signaling between potential mates among protozoa or green algae, or between the cells of fungi, or between the stigma and pollen of angiosperms, or the sperm and ovum in many animals involves a simple form of chemical recognition. In more advanced animals the signaling between the sexes is more complex, as in the case of the peafowl. The SMRS by means of which members of a species achieve positive assortative mating with considerable reliability can be described as a "signal-response reaction chain" involving both sexes or their cells. By "recognition" I imply nothing more than the sort of recognition involved in the recognition by a specific antibody of its antigen. I imply no act of choice.[19]

Such a system must exist for every biparental species. A little thought will show that positive assortative mating achieved in this way has an incidental and fortuitous consequence: it results in each species being "reproductively isolated" from all others! A little more thought will reveal that the whole class of heterogeneous bedfellows, the isolating mechanisms, about which so much fuss has been made since 1935, comprise nothing more than an artificial, man-made class of "effects."[11] They were none of them evolved as adaptations to protect the "integrity" of a species by preventing hybridization!

A crucial point to note is that the characters of an SMRS are closely adapted to function effectively in the normal or *recognized habitat* of the species. The signals involved may be auditory, visual, chemical, or tactile. Of these, man perceives visual signals, especially structural visual signals, most readily. Species which signal between the sexes in channels which man is not equipped to perceive, and whose signals are ephemeral, are generally hard to distinguish and are therefore called "sibling species."

When a small daughter population of a parental species becomes confined to a new environment to which it is poorly adapted, it may, of course, perish. However, if it survives, directional selection will adapt the population's descendants to the new circumstances, much as Darwin believed would happen. Since the characters of the specific-mate recognition system are adapted to the normal habitat of a species, they, too, will be shaped to be appropriate to the new conditions by natural selection.

When the process of adaptation is complete, the members of the daughter population will have become characterized by a new SMRS and recognized habitat. Should they ever again meet a population of the parental species, it is very likely that the members of the two populations will no longer exchange genes (i.e., hybridize). In other words, the daughter population will have speciated. Thus, a new species will have arisen as an incidental by-product of a small population adapting to the conditions of a new environment to which they had become restricted. As with Darwin's views, the recognition concept of species is entirely free of any commitment to teleology. [p. 275]

In accordance with this concept, members of a species can be seen to share a common specific-mate recognition system and a recognized habitat.

I arrived at this simpler and more effective concept of species after reconsidering species from first principles, which in this case is sex. I have only recently noticed how much the recognition concept has in common with Darwin's empirical view of species. Although Darwin's views were inadequate in detail, for he did not realize that species and varieties were qualitatively distinct, I, nevertheless, accept that they are philosophically ancestral to the recognition concept.

CONCLUSION

I am confident that we shall gain important and unexpected insights to the living world when we explore it through the recognition concept. Some will choose to avoid this path because it does not allow a purpose for man, nor a plan for life. My own strong conviction is that man will gain in stature if he is freed for the first time in history from a large number of ancient constraints on his mind, the "tyrannies of the past," as James Harvey Robinson once called them.[20] Surely this is a recipe for a new approach to living?

It is to me a sad commentary on our enlightenment, and on the effectiveness of our universities, to be pointing out a course which was first indicated 123 years ago by the quiet revolutionary whose death we are remembering tonight.

REFERENCES

1. Ellegard, A. 1958. *Darwin and the General Reader*. Gothenburg Studies in English. Vol. 8. Göteborg, Sweden: University of Göteburg, p. 201.
2. Ibid., p. 203.
3. Ibid., p. 14.
4. Ibid., p. 207.
5. Ibid.
6. Eiseley, L. 1979. *Darwin and the Mysterious Mr. X*. New York: Harcourt Brace Jovanovich.
7. Ray, J. 1727. *The Wisdom of God Manifested in the Works of the Creation*. 9th ed. London: William and John Innys.
8. The possibility that man creates his gods in his own image seems to be less often considered than it should be. Xenophanes of Colophon (c. 570–480 B.C.) wrote, after noting the appearance in paintings of Ethiopian gods, "If cattle and horses . . . had hands . . . horses would draw the forms of the gods like horses, and cattle like cattle."
9. Skolimowski, H. 1974. Problems of rationality in biology. In *Studies in the Philosophy of Biology*, ed. F. J. Ayala and T. Dobzhansky, 205–24. London: Macmillan.
10. Hull, D. 1973. *Darwin and His Critics*. Cambridge, Mass.: Harvard University Press.

11. Williams, G. C. 1966. *Adaptation and Natural Selection.* Princeton, N.J.: Princeton University Press.
12. Dobzhansky, T. 1951. *Genetics and the Origin of Species.* 3d ed. New York: Columbia University Press.
13. Darwin, C. 1867. Letter to T. H. Huxley, 7 January 1867.
14. Dobzhansky, T. 1976. Organismic and Molecular Aspects of Species Formation. In *Molecular Evolution,* ed. F. J. Ayala, 95–105. Sunderland, Mass.: Sinauer Associates.
15. Paterson, H.E.H. 1973. Animal species studies. Pp. 31–36 in Animal and plant speciation studies in Western Australia, by H.E.H. Paterson and S. H. James. *Journal of the Royal Society of Western Australia* 56:31–43. [See this volume, Author's Preface.]
16. Paterson, H.E.H. 1978. More evidence against speciation by reinforcement. *South African Journal of Science* 74:369–71. [This volume, chap. 3.]
17. Paterson, H.E.H. 1980. A comment on "Mate Recognition Systems." *Evolution* 34:330–31. [This volume, chap. 4.]
18. Paterson, H.E.H. 1981. The continuing search for the unknown and unknowable: A critique of contemporary ideas on speciation. *South African Journal of Science* 77:113–19. [This volume, chap. 5.]
19. Paterson, H.E.H. 1982. Perspective on speciation by reinforcement. *South African Journal of Science* 78:53–57. [This volume, chap. 8.]
20. Robinson, J. H. 1926. *The Mind in the Making.* London: Jonathan Cape.

10 The Recognition Concept of Species: Macnamara Interviews Paterson

Paterson, H.E.H., and M. Macnamara. 1984. The recognition concept of species (Michael Macnamara interviews H.E.H. Paterson). *South African Journal of Science* 80:312–18.

Michael Macnamara conceived the idea of this interview and received Graham Baker's approval before I heard of it. When it occurred, we both enjoyed the experience. The transcription to a manuscript was quick and smooth.

All considered, we thought the experiment a success, and I believe the publication has been helpful to others in explicating a number of points of recognition concept theory.

MM: Professor Paterson, when I compare current concepts of biological species, I find your "recognition concept" of species very thought provoking. Your ideas appear to be bringing about a shift in biologists' attitudes to the nature of species. I must say, however, that I am not sure how best to come to terms with your notions. Indeed, I have the impression that even certain leading biologists have not really grasped your meaning.

HP: Suppose you state the problems you consider most troublesome or significant.

MM: First and foremost: like several biologists with whom I have talked, I should like you to show clearly that your recognition concept of species is significantly different from the long-standing biological species concept or, as you prefer to call it, the "isolation concept."

HP: I can show immediately how radical the difference between these two concepts of species is by comparing them in terms of relational definition and with reference to the allopatric type of speciation.

The isolation concept is, as Mayr[1] has often emphasized, a *relational* concept: "Species are groups of interbreeding natural populations that are reproductively isolated from other such groups." Mayr thus defines species in relation to other species. He sees this relationality as an advantage, but it is, in fact, a fatal weakness. The reproductive isolation is said to be brought about by the action of ad hoc characters, the so-called isolating mechanisms. Now, Mayr is an ardent supporter of the mode of speciation known as allopatric (or geographical) speciation, i.e., speciation which occurs in total isolation, as on an oceanic island. But how, then, are *ad hoc* mechanisms (adaptations) presumed to arise in a situation of allopatry? This query exposes the major flaw in the isolation concept.

The recognition concept, on the other hand, is free of this defect because it is *not* a relational concept: "A species is that most inclusive population of individual, biparental organisms which share a common fertilization system."[2] No one can doubt that every biparental organism has adaptations which facilitate the occurrence of fertilization under the conditions imposed by its way of life and normal environment. Note that, in the case of the recognition concept, it is these fertilization mechanisms which determine the limit of the species' gene pool, i.e., its "field for gene recombination" as Carson has aptly called it. In the case of the isolation concept, it is the isolating mechanisms that are supposed to delimit the species' gene pool.

Thus, if we adopt the recognition concept, we can dispense with a whole class of adaptations, the isolating mechanisms, which have since 1937 been so much a feature of books on species and speciation.

MM: Why have you thought it necessary to rename the biological species concept the "isolation" one? Is this multiplication of names conceptually justifiable?

HP: The term "biological species concept" is not very descriptive. It is too wide: all genetical concepts of species are equally "biological." Moreover, as used by all leading authors, the term conflates two different genetical concepts of species. When I noticed this and duly separated the concepts, it became appropriate to dispense with the old, vague term and replace it with the two descriptive names I now use. These new terms are advantageous in identifying clearly the central property of each concept. The isolation concept merges with the biological species concept when the latter term is strictly used.

And now I have a pressing question for you as a philosopher. I have often been told that the recognition concept is not really new, that it involves nothing but a semantic difference. What am I to do with such a statement—ask what exactly the critics mean by "just a matter of semantics"?

MM: You would, I think, be justified in doing that. In its technical sense, the word "semantics" means that branch of semiotics (the theory of signs) which studies signs in relation to what, or how, they signify. (The other branches are syntactics, which studies signs formally and independently of interpretation, and pragmatics, which studies the purposes and effects of meaningful utterances.) But when people speak about "*just* semantics," presumably they are intimating that a given distinction drawn between certain items is a purely verbal one, rather than a conceptually or empirically significant one. We shall have to hypothesize about what they have in mind here. They may have in mind that some kind of linguistic sleight-of-hand has occurred or that a purportedly novel definition amounts to no more than the restatement of an existing definition in another terminology; or they may be thinking of the seemingly analogous situation in which heads and tails look different but are actually two sides of the same coin or perhaps of the stock semiotic example in which "the evening star" and "the morning star" are different names, but both refer to the same physical object, the planet Venus. Some of these metaphors and analogies could do with closer scrutiny. For example, if any of your critics are appealing to the Venus example, then they might notice that, as the logician Frege intimated in distinguishing between the "sense" and the "reference" of terms, the statement "Venus is Venus" is [p. 313] trivial, but the statement "The morning star is the evening star" certainly is not: the point is that the component phrases, though having the same referent, have different senses. Really, I think it is up to the critics to clarify the meaning of their criticism and to spell out their grounds for it. In any event, the "just semantics" criticism is rebutted if a new notion is shown to be conceptually distinct from the old one, or if the new and old concepts are incompatible in one or more crucial respects, or if the new notion is conceptually prior to the old one.

I now want to put my previous question to you in these terms: Do we really require yet another concept of species—has not the isolation concept stood the test of time?

HP: I do not believe it has stood the test of time. Surely one can justify the retention of a concept only if it is consistent with the principal facts? To take one example: the isolation concept is, as I intimated just now, not compatible with the only unquestionable mode of speciation, namely, the allopatric mode. Well, to my mind this is a fatal shortcoming.

MM: I am aware that you have also criticized the isolation concept because it involves the notion of group selection. But why do you object to this involvement?

HP: My cautious avoidance of the notion of group selection is part of my general approach to understanding the nature of species and speciation. In reaction to what I regard as the unjustified invocation of purely hypothetical principles by many workers in evolutionary biology, I have aimed to invoke only those principles which have a sound empirical basis. I believe that the foundations of my work will then be as reliable as they can be.

Even Wilson,[3] always a strong advocate of the concept of group selection, has recently reviewed the present status of the concept and has seen fit to conclude cautiously: "When faced with the difficult task of determining a character's function (if any), benefits to the group cannot categorically be excluded. The prevalence of such adaptations in nature remains to be seen because, despite the blast of words, group selection remains a largely unexplored concept whose important variables have only recently been identified and whose predictions are only beginning to be tested rigorously." I am thus inclined to avoid invoking the notion of group selection unless a given set of observations cannot be explained in terms of the well-established individual-selection model. I—and many others—will require convincing that any adaptation evolves "for the good of the species" rather than the "good of the individual."

In particular, I am reluctant to invoke the idea of group selection when the aim is to understand the nature of species, or the evolution of sex—a crucial problem in evolution. Thus, I avoid subscribing to statements like: "It is the function of isolating mechanisms to prevent such a breakdown and to protect the *integrity of the genetic system of species*" (ref. 1, p. 109; my italics). It follows from Mayr's sentence that isolating mechanisms are group adaptations which have evolved for "the good of the species," under group selection, rather than for the individual selection. We may note in passing that the usual model of speciation which involves natural selection for reproductive isolation, namely the reinforcement model, involves only individual selection!

Another common point of view which involves the notion of group selection is Dobzhansky's "adaptive polymorphism." Genetic polymorphism of the classical kind (heterozygote advantage), such as the well-known example of hemoglobin A and hemoglobin S in man, is clearly a group adaptation, if it is really an adaptation at all: it must have evolved for the good of the group, not the individual. But to me it seems likely that such polymorphisms arise fortuitously under individual selection and should not be called adaptive polymorphisms, but simply polymorphisms. Group functions should therefore not be sought for them.

This is not just a pedantic matter of definition; whether or not group selection occurs is of great importance when we seek to understand the nature of species—which means, ultimately, comprehending the nature of man.

At this point I have a question of a different kind for you. It seems to me that the isolation concept, in contrast to the recognition one, is teleological; but I am rather hesitant to use that word. What exactly does "teleological" mean to you philosophers?

MM: Yes, it is a rather slippery term: its meaning depends on field and on context—on whether the field is theology, metaphysics, ethics, epistemology, psychology, or biology, and then on what the context within that particular area is. Discussion of teleology goes back to Aristotle's introduction of his "doctrine of four causes." (The word "cause" has a wider meaning in the Aristotelian context than it has in an everyday one.) According to this doctrine, a thing has four kinds of causes, the material, formal, efficient, and final ones. Among these, only two need concern us here: his "efficient cause," which means what we signify nowadays by the term "cause" (i.e., "what antecedently or concomitantly produces an effect") and his "final cause," which is the goal or end (Greek *teleos*) toward which something tends naturally to develop. Aristotle's thinking about nature was indeed highly teleological. The medieval Scholastics in turn adopted his doctrine of causes and regarded all things as tending toward an end by "natural appetite." In contemporary times philosophers of science, notably Braithwaite,[4] have contrasted two main types of explanation, causal and teleological. To summarize: in general, an *explanation* is an answer to a why-question; specifically, a *causal* explanation has the backward-looking, causal form "because (of) . . . ," e.g., "The cat jumped because the dog nipped its tail"; and a *teleological* explanation has the forward-looking, goal-directed form "in order that . . . ," e.g., "The cat pawed at the doorknob in order that I should let it out." A teleological explanation thus consists in specifying a goal toward the attainment of which an event or activity is a means.

What is the status of teleological explanations? Are they scientifically respectable? Well, they are genuine accounts, if the criterion of authentic explanation is giving the questioner some appreciable degree of intellectual satisfaction; moreover, they are very pertinent in the case of sophisticated, intentional items of human behavior. Then why do many scientists tend to be chary of teleological accounts? I suppose there are at least three reasons. First, an "historical" reason: in a teleological explanation, according to which an event or action is said to be goal directed, the word "directed" implies a direction, which in itself is not an objectionable notion; but if it also implies—though it need not—a director, say a mystical First Cause or Deity, then a positivistically inclined scientist would, on applying the principle of economy of explanation (Occam's Razor), reject this explanation; he would want no talk of supernumerary metaphysical entities. Second, in a causal explanation something is comfortingly explained by a cause which either precedes or is simultaneous with it, whereas in a teleological explanation something is perhaps less comfortingly explained as being "causally" related to a remote goal in the future—sometimes in the very remote future. (This is the "puzzle of future reference.") Third, quite a [p. 314] number of scientists are mechanistically inclined, and such scientists seem bent on reducing all teleological explanations—even explanations of sophisticated actions in terms of conscious human intentions—to causal explanations couched in behavioral, or even physicochemical terms.

Now, while on the subject of views and terms, I want to pose two not unrelated questions: (1) You frequently invoke George Williams' view as embodied in his book, *Adaptation and Natural Selection*.[5] Why do you consider this book so important? (2) "Conflation" seems to be one of your favorite terms. In ordinary usage, to conflate is to fuse together or blend, for instance to blend two variant texts into one. And philosophers use the term in various contexts. But what exactly do you mean by conflation in the context of this interview?

HP: In reply to your first question, about Williams: I emphasize his book because I found that it revolutionized my thinking on evolution more than any other postwar text on the subject. The author wrote[5] (p. 258): "One aim of this book is to convince the reader that an understanding of the general nature of adaptation is important and that its study requires a more rigorously disciplined treatment than it ordinarily receives."

In particular, Williams made a crucial distinction between "function" and "effect." Imagine an invading Martian peering through a window at a writer whose papers are disturbed by a willful breeze. The Martian sees him solve the problem by taking out his pocket-watch and using it to restrain the sheets. Think of the problem facing the Martian when, having managed to get hold of the watch, he then tries to work out the rationale of its design while believing its function is that of paper weight! Williams calls the watch's effectiveness as a paper weight a fortuitous *effect*; the watch was originally shaped by man to serve the *function* of measuring time. The structure of the watch will be grasped by the Martian only when he realizes what it was designed to do, i.e., what its function is.

This story illustrates the sort of problem with which Williams himself grappled. Most authors are careless when speaking of adaptations and when assigning functions. As often as not, effects are called functions and, when this is done, a reader will not understand the role of adaptations, and how they arose. Let us now consider how authors such as Mayr see isolating mechanisms. Mayr, in the quotation given under an earlier question, states clearly that the function of isolating mechanisms is to isolate a species reproductively. I am wholly unconvinced. As I have often said, the postmating "mechanisms" (sterility, etc.) cannot have been selected to serve the function of isolating. Even Darwin[6] (p. 245) was aware of this. The sterility existing between some species is clearly an effect, not the function of the relevant character. When looked at closely, even the premating isolating "mechanisms" seem most likely to have been shaped by natural selection to serve quite other functions, and they isolate species as a fortuitous consequence; i.e., with respect to their isolating role, they are effects. Notice that this viewpoint is incompatible with many of the statements made by leading authors who support the isolation concept. Inconsistency is forced on these writers. For example, in order to accommodate his most favored model of speciation, namely, geographical speciation, Mayr proposes that the isolating mechanisms arise as incidental by-products of adaptation to a new environment. But this means that all his statements about isolating mechanisms being ad hoc characters, or about their function being to isolate, are wrong. There is certainly a price to pay for taking this "let-out." On p. 252 of his book[5] Williams wrote, "It [the species] is not an adapted unit and there are no mechanisms that function for the survival of the species."

Williams' book could eliminate the sloppy writing that has done such harm to evolutionary biology, provided its recommendations were clearly understood and followed. The distinction between functions and effects was also drawn later by Gould and Lewontin, in 1979.[20] They used an architectural analogy (the spandrels of San Marco), though without mentioning Williams' book. This is interesting because Lewontin actually reviewed Williams' book in *Science* (1966).[7] However, considering the points he emphasized in his review, it is evident that his original interest in the book did not include an appreciation of Williams' major insight!

Now, in reply to your second question, about conflation: I defined conflation, as I use it, near the start of my 1981 paper [chap. 5].[8] Following Wittgenstein, I said that we have conflation when an author includes two distinct concepts under one heading (e.g., two different concepts of species under the same name, "species"). He can then glide imperceptibly from discussing species under one concept to talking about them under the other. This glide generates a subtle kind of nonsense which may not be easily detected. Usually the conflation is not done consciously, but occurs due to the author's failure to appreciate the difference between the two concepts. The way authors conflate the taxonomic and the two genetical concepts of species is startling. They may flit from one to another, and even to the third, within just three lines of text. Here again we have a major problem in the literature and in discussions between evolutionists. Moreover, conflation makes the testing of hypotheses very difficult. White[9] wrote a book on speciation models without once defining clearly what he meant by species. It is not surprising, then, that he switched from one concept to another to suit his prevailing need. In such cases, what is convenient for the author is fatal to clarity of understanding.

I often wonder why it is that the flaws I have pointed out in the argument for the isolation concept have not been identified by philosophers of science who specialize in evolution. I think philosophers of biology have an important duty to keep us biological scientists on the right track. Is it not their duty to tell us when we are conflating, when we are being grossly teleological, and when we are avoiding issues? What, in short, do you see as the role of the philosopher of science?

MM: I should first say that, though interested in biology, I certainly do not claim to be either a biologist or a philosopher of biology. Until recently, in fact, my interest in the philosophy of science was focused pretty exclusively on the physical sciences.

You ask what the role of the philosopher of science is. How that role is seen depends on what one's larger conception of the nature and task of philosophy as a whole is, and there are several different such conceptions; examples are the logical-positivist, conceptual-analytical, existentialist, phenomenological, and Marxist ones. To state one contrast between views in terms of the envisioned status of science: the positivistically inclined empiricist philosophers, like Bertrand Russell and the members of the Vienna

Circle, held science in very high esteem, whereas an existentialist like Jean-Paul Sartre was convinced that it is not science at all, but literature-oriented philosophy, that connects most directly with the primacy of experience. Anyhow, I think some of the conceptual analysts had a good point when they drew a broad working distinction in orders between the activity of the special scientist, who studies the [p. 315] world and makes first-order statements about it, and the task of the philosopher of science, who makes second-order statements about the scientist's statements about the world. Of course, one cannot press the distinction too far, for the two parties do—and, I think, should—sometimes drift into each other's territories, but there does remain this broad difference in their professional training and concerns. I should say the philosopher of science is concerned largely with matters such as the analysis and clarification of basic scientific concepts, terms, and definitions; the validity of deductive scientific arguments; the identification and elucidation of key scientific postulates, axioms, and presuppositions; the nature of scientific principles, methods, theories, and models; the justifiability of scientific inductions; the doctrines (yes) that underlie science; the problem of the reducibility of one science to another; and the relations between the various special sciences and between science and other fields of human activity.

You also ask why philosophers of biology who specialize in evolution have not spotted flaws in the isolation concept. I do not know why.

As regards your question about the "duty" of the philosopher of biology: you exemplified your query in terms of method (conflation), of underlying theory or doctrine (teleology), and of stand (facing or dodging issues). Yes, I think the philosophers ought to feed back to the biologist on all these matters.

Having just mentioned models, I should like to raise the following question: You and your colleagues (Centner, Lambert) have used simple mathematical and computer models to study aspects of evolutionary theory: but just what, and how much, do such studies tell?

HP: This is a good question, because computer modeling is now common practice in population and evolutionary genetics and in ecology. I believe models can be useful and informative, but much depends on the circumstances. We have only very modest skills in this field, and our models are simple. But, though simplified, they are often better than verbal models because of the well-known problem with verbal scenarios: they reflect our preconceptions. Mathematical models are less susceptible to such reflection, always supposing that they are not deliberately made to mirror preconceptions.

Let me illustrate my remarks by discussing what is likely to happen when a population buds off a small daughter population, which then diverges in a distinct habitat but becomes sympatric again before it has acquired species status. Many authors have postulated that, when individuals from the two populations cross, their offspring will be less fit than the members of either of the parental groups. Now, assuming that mating occurs at random, deterministic algebra shows clearly that a metastable equilibrium will arise if the populations are present in equal numbers. However, should any disturbance of the proportions occur by chance, or for any other reason, then the less frequent causative allele or chromosome arrangement will be eliminated at a rate determined by its disadvantage. If the hybrids are totally sterile, then the less abundant population will be eliminated. This expectation is commonplace to population geneticists. Curiously enough, however, deterministic algebra seems not to have been used for this purpose before my 1978 paper[10] appeared.

In that paper [chap. 3], I used this model after having presented a verbal scenario of what the well-known evolutionary geneticist, Ayala, predicted the outcome of such a situation would be. Nor was he alone. Other leading authors, including the great mathematical evolutionary geneticist Sir Ronald Fisher[11] (p. 145), did the same. When authors model such a situation verbally, the outcome is predetermined by their expectations: the meeting of the populations leads to the formation of disadvantageous hybrids; natural selection, acting against these, leads to the evolution of mechanisms—isolating mechanisms—to prevent such wastage; and, as these mechanisms evolve, hybridization will decrease and speciation will be completed. Now this verbal scenario does not accord with the simple mathematical model, which predicts that natural selection will rather tend to remove the cause of the heterozygote disadvantage. However, the verbal model sounds so convincing that nobody has questioned it for over fifty years. Even when the situation had been modeled on computers by Crosby[12] and Wilson[21] and the prediction of the algebraic model was confirmed, the authors sought to shore up their model till the expected result was obtained. Similarly, when this situation was studied experimentally (e.g., Crossley,[13] Koopman[14]) using *Drosophila*, the reinforcement hypothesis was judged to have been strikingly verified (Dobzhansky,[15] p. 209). What escaped the experimenters' attention, due to their ignoring the mathematical formulation, was that their experimental design precluded the outcome expected from the mathematical model. This is discussed in detail in my 1978 paper;[10] Harper and Lambert[16] have recently (1983) repeated the *Drosophila* experiments, but they took care not to block the selective removal of the cause of the disadvantage, and their experiments—not surprisingly—confirm the predictions of the mathematical model. This long story illustrates one advantage of mathematical modeling over verbal modeling: it is more objective. And even mathematically inclined workers can be seriously led astray by verbal formulations because of the unsuspected influence of preconceptions.

When mathematical models yield answers that are unpalatable to an author, it is of course always possible to incorporate additional conditions which will, with luck, produce the desired answer (cf. Crosby,[12] p. 259). But such ad hoc shoring up should be regarded with considerable skepticism, and the following question must be asked: If you know what the outcome ought to be, why bother to model the situation?

To sum up: mathematical models, when used carefully, can be more objective than verbal formulations. But one should always look closely at the premises of both a verbal and a mathematical model to find whether they are unrealistic or incomplete. Models, then, are valuable within limits. However, they should never be treated as empirical data. None of the mathematical or verbal models for sympatric speciation, for example, can be considered as more than hypothetical until empirical support from nature is obtained. That is why I have suggested that they be called Class II models[8] (as opposed to empirically supported Class I models).

MM: You referred just now to the budding-off of small populations. But on what grounds do you believe that speciation occurs in small rather than large populations?

HP: Another of the constraints which I impose on my theorizing is to argue from an empirically supported position whenever possible. We know, with quite a high level of certainty, that speciation can occur in small populations; on the other hand, we have only hypothetical models to support the thesis that speciation occurs in large populations which are distributed widely. Now what is the empirical support for the thesis that speciation occurs in the small populations to which I refer? Quite commonly, we find two [p. 316] closely related species which have alternative gene arrangements (e.g., translocations, or pericentric inversions) fixed in the two populations. In such cases, we can be sure that the derived arrangement was not spread through the population under selection, for the heterozygotes of such arrangements are mechanically disadvantaged during meiosis. Short of invoking rare and improbable mechanisms, such as meiotic drive, the only remaining possibility is the mechanism of stochastic fixation in a small population.

Of course, hypothetical models for speciation in big, widespread populations can be put forward. But it must be remembered that this can raise problems. In short, the important thing to keep in mind is that we do not have empirical support for the thesis that speciation occurs in big populations, whereas we do have such support for the claim that speciation takes place in small populations. Consequently, I prefer to restrict myself to arguing from the position that is better supported.

MM: You have indicated that, in common with Ernst Mayr and others, you have a strong belief in the occurrence of allopatric (or geographical) speciation. But are you as convinced as Mayr seems to be that speciation invariably occurs in this way?

HP: I shall follow much the same reasoning in answering this question as I did for the last one. We have very many examples from islands, lakes, and mountains that constitute evidence for allopatric (or geographical) speciation. (As both Dobzhansky and White have said, no one doubts that such speciation occurs.) For this reason I regard geographical speciation as empirically supported, and I have assigned it alone to Class I (i.e., empirically supported) speciation models. All other models, being unsupported empirically, must be regarded as hypothetical. Of course, many models have been contrived with the isolation concept in mind. But, since this concept is unacceptable, many of the models of speciation designed from it must fall away. I prefer not to build theory on insecure foundations; therefore I invoke only allopatric speciation.

It should be emphasized that the recognition concept imposes different strictures from those enforced by the isolation concept on any model of speciation. Workers with a predilection for designing models of sympatric speciation will have to start again in the case of the recognition concept; the old models, including "instant speciation" by autopolyploidy, will not be appropriate.

Thus I should say in answer to your question that I know of no convincing evidence for any other model. However, this answer does need a rider. It should be realized that, because Mayr was obliged to rely on the notion of pleiotropic change to isolating mechanisms during speciation in allopatry, his ideas on how "allopatric speciation" occurs are quite distinct from mine. So, allopatric speciation should not be regarded as only one proprietary process.

Although I believe that the pleiotropy which Mayr invokes may play an important role in speciation, just as stochastic events may, I also believe that the most common event in the process involves the "shaping," by natural selection, of characters of the fertilization system to the conditions of a new environment. Even characters that are affected by pleiotropy or drift will ultimately have to survive testing by natural selection.

MM: You have mentioned one or two implications of your recognition concept for population biology. But how comprehensively does your concept affect the several regions of this discipline?

HP: The recognition concept affects nearly all aspects of population biology, at least indirectly. Sir Charles Lyell emphasized the ramifications of species concepts when he said in the early thirties of the nineteenth century, "The ordinary naturalist is not sufficiently aware that, when dogmatizing on what species are, he is grappling with the whole question of the organic world and its connection with a time past and with man."

Within evolutionary biology, the recognition concept raises quite distinct expectations as to how speciation occurs, and it is, as I intimated earlier, free from teleological overtones. Also, it calls for nothing more than individual selection and it dispenses largely, at least, with the need to invoke sexual selection. In the field of behavior these simplifications have important consequences: courtship behavior is seen wholly in terms of securing effective fertilization; there is no need to account for such behavior in the language of isolating mechanisms. In ecology, the change in species concepts has an important bearing on competition theory, the understanding of the niche, character displacement, and the evolution of ecosystems. In paleontology, the recognition concept provides a sound theoretical model from which that pattern of variation in time which is known as "punctuated equilibria" is derivable. The concept makes predictions on the kinds of organisms which would most effectively test the ideas of punctuated equilibria, the effect hypothesis, the Red Queen hypothesis, and so on. In anthropology, the concept provides new predictions on how *Homo sapiens* evolved. It has eliminated the difficulty of accounting for the nature of man's isolating mechanisms, and how they could have arisen, and it replaces the refractory notion of such mechanisms with the convincing alternative picture of a fertilization system. The philosophical implications for man are important. I repeat that the recognition concept, unlike the isolation concept, is free from teleological overtones; I might add now that it provides Darwin's thesis of nonteleological speciation with the species concept that it has lacked. The implications of the recognition concept are thus considerable.

MM: I wonder, though, if all this is not a bit academic and recondite. Could not the working biologist simply ignore your distinctions between these two species concepts?

HP: That is a difficult question. I cannot answer for others, but to me there are no such distinct things as "academic" and "nonacademic" science. I worked for many years in applied entomology, so I am not the sort of scientist whose experience is restricted to the ivory tower. I do not now do anything different from what I did then: I am motivated in the same way, whether I am working on the epidemiology of malaria or striving to understand the nature of species. Fundamentally, I am interested in comprehending the diversity of life on earth. Methodologically speaking, I do not believe that it is ever justifiable, in science, to cut corners. Much of applied science, ecology, and, indeed, work in other branches of biology is worthless, or of greatly diminished value, because of workers' lack of care in determining the identity of the species being studied. To exemplify, the existence and limits of sibling (cryptic) species are ignored as merely academic, with a consequent loss of insight that is often vital. In particular, I do not believe in drawing a distinction between field biologists and theoretical biologists. Field workers are scientists only if [p. 317] they are testing questions, and it is theory that generates questions. Theoreticians who do not know organisms at first hand, and intimately, are liable to set nonsensical premises to their models. My reply is that each scientist must answer the question of concept choice for himself or herself. I suppose it is of no consequence which of the two concepts a worker puts to use if he fails to grasp what differences his choice entails. However, his choice certainly has import for me if I am examining his work to determine whether it is of value for my studies. I first test his work for quality. If it fails, I simply ignore it; otherwise I may even use it in my teaching as a dreadful warning to students. As regards this whole topic of species, there are many scientists who would be perfectly happy to remain in the pre-Darwinian era, but then their aims are neither to understand evolution nor to benefit from the insights provided by it. To conclude this answer, let me summarize my attitude by saying that, in my experience, the most practical education is the most theoretical, which is another way of saying that theory does matter!

MM: You have drawn several distinctions between the alternative concepts—isolation and recognition. But can these opposing concepts be subjected to a crucial, experimental test?

HP: Yes, it *ought* to be possible. But man has ever so subtle a mind. Many a test does expose a weakness, yet this is promptly taken care of by what I call "ad hoc shoring up," i.e., the patching up of an hypothesis to the extent that it begins to resemble some product of Heath Robinson's fertile mind. As regards crucial criteria, for me two facts effectively falsify the isolation concept: the fact that allopatric speciation is the only empirically established mode of speciation and the fact that allopatric speciation cannot account for the origin of isolating mechanisms if the latter are regarded as *ad hoc* devices or adaptations for protecting the integrity of species. However, other biologists appear to be more accommodating than I.

The recognition concept, in its turn, would be falsified effectively if compelling evidence could be adduced that the courtship behavior of, say, the three-spined stickleback had evolved to prevent hybridization with the ten-spined stickleback or that the species-specific male genitalia of *Sarcophaga haemorrhoidalis* had evolved to prevent hybridization with other closely related congenerics. I would also concede that the concept had been falsified if it could be proved that a particular species, and its species-specific fertilization system, had arisen in true sympatry. A little thought will indicate that these conditions are difficult to satisfy.

But how do you philosophers view the ad hoc shoring up of a concept which is under siege? Did Popper not have something to say about it?

MM: Yes, Popper[17] (p. 83) sounded a warning against the uncritical introduction of an ad hoc hypothesis to bolster a questionable theory. He did, however, allow that there are acceptable as well as unacceptable ad hoc hypotheses. His criterion was that an acceptable ad hoc hypothesis is one that increases the testability or falsifiability of a system. Taking examples from physics, he cited the Fitzgerald-Lorentz contraction, which he said had no falsifiable consequences, as an unacceptable ad hoc hypothesis (when it is viewed as squaring the results of the early Michelson-Morley experiment, rather than as a consequence of later relativity theory); and he named as acceptable the Pauli exclusion principle (according to which no two electrons in a neutral atom can have the same set of four quantum numbers). Later, philosophers like Hempel[18] (pp. 28–30) used a distinction between gainful *auxiliary*, and purely *ad hoc*, hypotheses. The innocent auxiliary hypothesis or assumption, which is the rule rather than the exception in science, fills in or clears up the steps in the inference of an implication from the main hypothesis or theory, whereas the suspect ad hoc hypothesis is added in an attempt to save a theory "come hell or high water." A good example of the latter arose over the early theory of heat, according to which the combustion of metals involves the escape of a substance called phlogiston. This theory was generally abandoned when Lavoisier showed that the end product of combustion has, in actuality, an increased weight; but some determined champions of the phlogiston theory ventured to rescue it by advancing the ad hoc hypothesis that phlogiston has negative weight. Of course, deciding whether something is an innocent auxiliary or a guilty ad hoc hypothesis is always much easier from a position of hindsight.

But to answer your question in more general terms: on the one hand, it must be conceded that some of our most august theories or systems, like quantum mechanics in the exact sciences, are patchy—the latter is a veritable tartan of classical, relativistic, and wave mechanics, of physical and mathematical models, of postulates and analogies; on the other hand, no genuine philosopher of science could countenance the introduction of an ad hoc hypothesis for scientifically irrelevant reasons such as the last-ditch rescue of an ideology, social expediency, psychological perseveration, or plain bloody-mindedness.

To return to specifically biological matters: Have you an example of an attempt at the ad hoc shoring up of the isolation concept of species itself?

HP: Yes, indeed. Both Dobzhansky and Mayr stated, at one time or another, that the ubiquity of allopatric speciation did constitute a problem for the isolation concept, but they both tended to play this down. Dobzhansky[19] (p. 376), discussing allopatric speciation and speciation by reinforcement (i.e., speciation under selection for reproductive isolation), wrote: "These two hypotheses are not mutually exclusive. Needless disputes have arisen because they were mistakenly treated as alternatives." His attempted solution was purely *assertive*, for he provided no evidence in support of his surprising claim. Mayr's try at shoring up was to state that the new isolating mechanisms arose in allopatry as a by-product of a population adapting to the new conditions in which it had become isolated. This, in itself, is not a solution, for any character which arises in allopatry by this means cannot be called an isolating mechanism; it cannot be called an *ad hoc* character that evolved to protect the integrity of the species; it is clearly an effect, in Williams' sense. If the latter stand is taken, much of what Mayr and Dobzhansky have written about species, their significance, and the nature of isolating mechanisms becomes meaningless. That is why I say that their shoring-up efforts are no real solution to the problems of the isolation concept.

MM: To sail on another tack now: I have heard botanical mutterings about the applicability or otherwise of your recognition concept. Do you maintain that the concept holds for plants as well as animals?

HP: Yes, it does. Although we often hear botanists say "plants are different," they do not specify in what respects. The fact of the matter is that the most important feature of biparental plants, in contrast to most animals, is due to the plants being sessile. Plants face much the same set of problems, in [p. 318] bringing about fertilization, as do sessile animals. For example, oysters and brown algae of intertidal rocks have very similar fertilization systems.

My first definition of species in terms of the recognition concept was framed as follows: "Members of a species share a common specific-mate recognition system." This definition applies pretty well to motile animals, but not to sessile organisms in which the specific-mate recognition system (SMRS) is not well developed. Well, I can—with reference also to your previous question—demonstrate an "auxiliary" adjustment (à la Hempel) by showing you how I modified the definition in order that it could apply to all biparental organisms: "A species is that most inclusive population of individual, biparental organisms which share a common fertilization system."[2]

MM: I believe botanists have also criticized the recognition concept for not accommodating organisms that have the asexual, i.e., uniparental, mode of reproduction.

HP: Certainly the recognition concept does not accommodate such organisms; but neither does the isolation concept, since both are genetic concepts. Two points should be made in response to this question. First, most multicellular, uniparental organisms are clearly derived from sexual ancestors, and I believe, like Darlington, that they often represent an "escape from hybridity." Second, since there are very few primarily asexual eukaryotic organisms, they should be regarded as a special but different case and could simply be handled under a taxonomic definition of species. But they are not the same thing genetically.

MM: I suggest that you might, at this point, profitably summarize your reasons for supporting the recognition concept *contra* the isolation one.

HP: I support the recognition concept because it accounts for nearly all the facts known to me in a simple, elegant, and intuitively satisfactory way. It is applicable to all biparental eukaryotes, and, in contrast to the isolation concept, it is compatible with the only empirically supported model of speciation, i.e., the allopatric one. Indeed, it elucidates how speciation in allopatry can occur, and it does not rely on invoking pleiotropy. It is not teleological; in fact it allows species to be seen as "effects." It leads to a number of important predictions for ecology and population biology as a whole, including anthropology. In contrast, the isolation concept presents such major difficulties that I believe it can no longer be espoused. There are too many inconsistencies in it.

MM: To conclude this interview: How does your recognition concept of species figure in the controversy between the advocates of phyletic gradualism in evolution and the proponents of the Eldredge-Gould pattern of punctuated equilibria?

HP: This controversy, between the supporters of the pattern of punctuated equilibria and those supporting the pattern of phyletic gradualism, has disappointed me and others because of the poor quality of the debate and the emotion with which it has been pursued. I am not keen to participate in it to any extent. However, I must confess that, if I ask myself the question, "Which of the two patterns would be predicted by the recognition concept?" I am quite sure I would expect the pattern of punctuated equilibria to be the choice. This is because the recognition concept takes into account the following features which are crucial in the debate on the two opposing patterns:

1. As I said earlier in reply to your question about small populations, it is evidently in small populations that speciation occurs.
2. Once evolved, a species is stabilized for several reasons, viz.: (a) its restriction to a habitat similar to the one in which it evolved— this factor keeps many adaptive characters under stabilizing selection; (b) the coadaptation of males and females for the signaling characters of the SMRS; and (c) the size of the population, after it has grown as a result of its release from directional selection (see ref. 2).

Thus, speciation in small populations makes it probable that the change from one character state to the next (i.e., punctuation) is unlikely to be detected in the fossil record. The stability of the population after speciation is the basis of the "equilibrium phase." The punctuation aspect of speciation could admittedly be accounted for under the isolation concept, but that model does not predict stability after speciation, whereas the recognition concept does.

MM: Professor Paterson, thank you for expounding and clarifying aspects of the recognition concept of species.

REFERENCES

1. Mayr, E. 1963. *Animal Species and Evolution.* Cambridge, Mass.: Belknap Press of Harvard University Press.
2. Paterson, H.E.H. 1985. The Recognition Concept of Species. In *Species and Speciation,* ed. E. S. Vrba, 21–29. Transvaal Museum Monograph No. 4. Pretoria: Transvaal Museum. [This volume, chap. 12.]
3. Wilson, D. S. 1983. The group selection controversy: History and current status. *Annual Review of Ecology and Systematics* 14:159–88.
4. Braithwaite, R. B. 1953. *Scientific Explanation* [chap. 10]. London: Cambridge University Press.
5. Williams, G. C. 1966. *Adaptation and Natural Selection.* Princeton, N.J.: Princeton University Press.
6. Darwin, C. 1859. *On the Origin of Species by Means of Natural Selection; or, the Preservation of Favoured Races in the Struggle for Life.* London: John Murray.
7. Lewontin, R. C. 1966. Review of *Adaptation and Natural Selection. Science* 152:338–39.
8. Paterson, H.E.H. 1981. The continuing search for the unknown and unknowable: A critique of contemporary ideas on speciation. *South African Journal of Science* 77:113–19. [This volume, chap. 5.]
9. White, M.J.D. 1978. *Modes of Speciation.* San Francisco: W. H. Freeman.
10. Paterson, H.E.H. 1978. More evidence against speciation by reinforcement. *South African Journal of Science* 74:369–71 [This volume, chap. 3.]
11. Fisher, R. A. 1958. *The Genetical Theory of Natural Selection.* 2d ed. New York: Dover Publications.
12. Crosby, J. L. 1970. The evolution of genetic discontinuity: Computer models of the selection of barriers to interbreeding between species. *Heredity* 25:253–97.
13. Crossley, S. A. 1974. Changes in mating behaviour produced by selection for ethological isolation between ebony and vestigial mutants of *Drosophila melanogaster. Evolution* 28:631–47.
14. Koopman, K. F. 1950. Natural selection for reproductive isolation between *Drosophila pseudoobscura* and *Drosophila persimilis. Evolution* 4:135–48.
15. Dobzhansky, T. 1951. *Genetics and the Origin of Species.* 3d ed. New York: Columbia University Press.
16. Harper, A. A., and D. M. Lambert. 1983. The population genetics of reinforcing selection. *Genetica* 62:15–23.
17. Popper, K. 1965. *Logic of Scientific Discovery.* Revised Torchbook ed. New York: Harper and Row.
18. Hempel, C. G. 1966. *Philosophy of Natural Science.* Englewood Cliffs, N.J.: Prentice-Hall.
19. Dobzhansky, T. 1970. *Genetics of the Evolutionary Process.* New York: Columbia University Press.
20. Gould, S. J., and R. C. Lewontin. 1979. The spandrels of San Marco and

the Panglossian paradigm: A critique of the adaptationist programme. *Proceedings of the Royal Society of London, B* 205:581–98.

21. Wilson, E. O. 1965. The Challenge from Related Species. Chap. 2 in *The Genetics of Colonizing Species*, ed. H. G. Baker and G. L. Stebbins, 7–27. New York: Academic Press.

11 Simulation of the Conditions Necessary for the Evolution of Species by Reinforcement

Lambert, D. M., M. R. Centner, and H.E.H. Paterson. 1984. Simulation of the conditions necessary for the evolution of species by reinforcement. *South African Journal of Science* 80:308–11.

Marc Centner and Dave Lambert were the actual authors of this paper. I was included, I suspect, because this extended my simple algebraic model of 1978 (chap. 3). It is interesting because it uses an empirically determined rate of change in the degree of positive assortative mating. This model should be compared with the later model proposed by Spencer, McArdle, and Lambert (1986, 1987).

Of some interest is the fact that Ernst Mayr was a referee for this study.

ABSTRACT

The contribution of the process of reinforcement to speciation requires that hybrids between two "semispecies" are of, at least, reduced viability. Workers in the field of speciation theory have generally not considered the fact that such a scenario is analogous to the population genetical description of the phenomenon of negative heterosis. We have formulated a model which describes the dynamic interaction between the evolution of positive assortative mating and the population genetical effects of negative heterosis. Our model requires the use of an empirically derived rate for the change in the degree of positive assortative mating. With, in our opinion, the most reliable estimate of this rate of change that we could obtain from the literature, our model strongly indicates that the elimination from the combined population of the "alleles" causing reduced hybrid viability is a more likely outcome than the evolution of two good species.

INTRODUCTION

There have been primarily two opposing views in biology as to how species arise. One theory maintains that populations of an original species diverge initially in allopatry. Upon secondarily overlapping, provided hybrids between the two are disadvantageous, natural selection causes the evolution of "premating isolating mechanisms" by selecting against those individuals which mismate with members of the opposite group. Eventually individuals of the two groups will no longer recognize each other as mates; i.e., "ethological isolation" is complete and the speciation event is concluded. This has commonly been called speciation by reinforcement.[1] This theory has been advocated by a long line of the most distinguished biologists, including Sir Ronald Fisher and Theodosius Dobzhansky. Today it occupies a central position in evolutionary biology. MacArthur and Wilson[2] have based much of their theorizing on this model, and Lewontin[3] in his influential book writes as if no other model of speciation exists.

In contrast, authors, beginning with Darwin himself, and including such well-known biologists as Ernst Mayr, Herman Muller, John Moore, and Wilson Stone, have placed more emphasis on divergence of both premating and postmating isolating mechanisms while populations are geographically separated.

In order to test the reinforcement model, it was our aim to simulate the forces acting on a population in the stages of complete and partial secondary overlap and hence to simulate the conditions under which natural selection could lead to the evolution of distinct species. Once these conditions were detailed it might be possible to estimate whether or not these conditions are likely to be found in natural populations.

Powerful, though indirect, evidence for the reinforcement model is the well-known laboratory selection experiments known to be capable of selecting for increased positive assortative mating [p. 309] between populations by selectively eliminating hybrid progeny.[4–9] However, Paterson[10] [chap. 3] has argued that much of the work in this area is flawed. Many authors have not considered that the reinforcement situation fits the population genetic model of negative heterosis. Where selection against hybrids is absolute, the model predicts the total elimination of the rarer genotype and, where the hybrids are of intermediate fitness, elimination of the rarer of the alleles causing the disadvantage. Ayala[11] has not considered this possibility and has maintained that under conditions of secondary overlap there are two possible outcomes. First, a single gene pool may come about because the populations hybridize readily without significant loss of fitness, or alternatively, two species may ultimately arise because natural selection favors the development of positive assortative mating between the populations. Paterson[10] [chap. 3] has argued that another alternative, elimination due to heterozygote disadvantage, needs to be considered. He has pointed out that the selection experiments mentioned previously have eliminated the possibility of extinction of the rarer population by artificially making the numbers of the two types equal in each discrete laboratory generation. Harper and Lambert[12] have presented laboratory evidence which shows that modified selection experiments do, in fact, lead to the elimination of one type when negative heterosis is allowed to act. This study also illustrates a tendency to eliminate one allele when hybrids have partial viability.

Wilson[13] and Bossert[14] argued that the outcome of conditions of secondary overlap would depend upon the interplay between conflicting possibilities. Wilson[13] commented: "The final act of speciation can, therefore, be viewed as a race between hybridization and (reproductive character) displacement."

An extension of the previous situation is that involving parapatric distributions. In the case where hybrids are of intermediate fitness, we would expect linkage groups carrying genes responsible for decreased fitness of hybrids to be selectively eliminated in the zone of sympatry. Key,[15] in his analysis of hybrid zones, emphasized this possibility, and Zaslavsky[16] discussed the nature of parapatric zones while considering possible elimination of the rarer form. A striking example of the selective elimination of linkage groups has been provided.[17]

This model attempts to examine the relationships between these two key processes, elimination of the cause of hybrid disadvantage and the formation of distinct species. We have done this by accepting that, since the process by which an individual recognizes a mate is genetically determined, selection on such a trait must be capable of causing a change in the process of specific-mate recognition. However, the process of elimination of the cause of the disadvantage of hybrids inevitably must also continue. Hence we have analyzed the change in these contrasting trends of two populations over discrete generations with varying conditions.

A GENERAL SIMULATION OF REINFORCEMENT

In contrast to other models,[14,18,19] we have not created a simulation in which a set of characters undergoes selection for increased premating isolation. We have identified those factors which lead to determining an overall intensity of selection for premating isolation. We show that the "intensity" is neither constant through time nor identical for either of the pure groups or the hybrids.

Since (assuming that some nonsterile hybrids survive) the outcome of any type of hybrid cross is essentially the same, i.e., half the offspring would be hybrids and half one or other or both parental group types, there is no particular pressure for hybrids to favor one or other parental group or the hybrid group as a source of potential mates. Thus, for this reason and also because it is difficult to predict the nature of the inheritance of the components of specific-mate recognition systems, we have chosen to assume that hybrids are nondiscriminating in their choice of mates.

However, for parental groups, nonassortative mating is a function of the compound probabilities of meeting, accepting, and being accepted by a potential mate. This, in turn, is a function (assuming total sympatry) of the relative numbers of the three groups and the behavioral mechanisms whereby individuals recognize and accept and are recognized by potential mates. Since intragroup mating is advantageous in this situation, the evolutionary cost and hence the intensity of selection for change can be equated to the probability of an individual mating with a nongroup member, the number of hybrid offspring produced thereby, and the selection coefficient applied to hybrid offspring.

Coupled to the intensity of selection, we include a constant, k, which essentially represents the contents of a "black box" which describes the interaction of those factors which contribute to the stability or potential for change in specific-mate recognition systems. Such factors might include the amount of variation, the coadaptation of signals and responses, and the complexity of the repertoire of courtship behavior. Our approach lays great stress on the fact that only empirical measurement of the outcome of the processes symbolized by this black box can provide a reasonable measure of the rate-limiting step in the reinforcement process. Section 1 of the appendix shows the formal derivation of the different equations describing the intensity of selection for change in the mate recognition system. The equations calculating the genotype distribution in the next generation are derived in section 2 of the appendix.

In the formulation of this simulation the unstated assumption (implicit in intraspecific population genetical theory) is that there is no ecological differentiation between the two "homozygous" groups. Competitive elimination of one group by another presents a trivial solution in the context of this problem.

One possibility which can affect the outcome of the simulation is where it is postulated that the population numbers of the two groups are totally independent of each other. Given that these hypothetical organisms have sufficient reproductive capacity to make good the losses incurred through hybridization, then a situation may ultimately arise where their numbers, and hence the relative frequencies, could remain constant and therefore allow reinforcement to take place. We know of no situation in which these conditions are realistic when applied to "incipient species" and therefore do not take this type of situation seriously. [p. 310]

RESULTS

It was the purpose of this simulation to determine the minimum value of k which caused complete divergence in the specific-mate recognition systems of the two groups before the process of elimination was complete. Figure 1 shows the effects of a number of k values on these two interacting processes. By varying all the parameters of the simulation, a k value of 0.15 was shown to be the minimum value necessary to just cause successful reinforcement.

In applying our modeling technique in order to determine the likelihood of reinforcement occurring in real populations, it is of crucial importance to estimate values of k that reflect the rates of change in the systems of specific-mate recognition of organisms. The only study, of all the selection experiments previously mentioned, that allows a direct estimate of k is that of Crossley.[9] In our calculations we have taken cognizance of the fact that equal numbers of the mutants of *Drosophila melanogaster*, ebony and vestigial, were used and that no hybrids were saved. A number of assumptions have been made which incline to result in an overestimate of k, but this tends to strengthen the argument below. Section 3 of the appendix fully describes our calculations, which result in an estimated k value of 0.016.

Using this value of k in our simulation (Fig. 1) indicates that rapid extinction of the rarer population rather than reinforcement of positive assortative mating is the likely outcome.

By considering hybrids with a selective value of unity we have modeled the extreme case in which either hybrids are sterile or no viable offspring are produced. However, any reduction in the value of s decreases the "cost" of mismating and hence the rate of change of d. Figure 2 shows that under similar circumstances a lower s value results in a smaller change over a given number of generations.

While our stated intention was to model both sympatric and parapatric cases, the latter turns out to be a trivial one. For the reasons outlined by Paterson,[10,20] there is little likelihood of individuals with

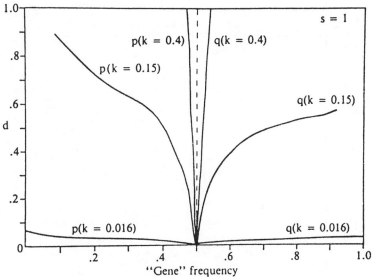

Figure 1. Under conditions of complete hybrid sterility, *k* values of 0.4, 0.15, and 0.016 are shown. Curves in the left half of the graph represent genes tending to extinction, and those on the right represent those tending to fixation. The ordinate axis represents the divergence in systems of recognition. *d* = 0 for each population represents random mating; *d* = 1 for one or other population represents complete positive assortative mating. Any curve intersecting the vertical axis represents a situation in which reinforcement has not occurred, and, similarly, intersection of the horizontal axis where *d* = 1 represents the formation of species by reinforcement. In order to give maximum opportunity for reinforcement to occur, numbers of each pure group were made almost equal. "Gene frequency" is defined in the text and the appendix.

an altered specific-mate recognition system spreading away from the parapatric zone. On the other hand, migration into the parapatric zone would tend to dilute any change that has already taken place. In these circumstances, a stable equilibrium with some degree of divergence in *d* could occur in the parapatric zone. This, however, would not lead to the ultimate reproductive isolation of the two parental groups. Reported cases of partial reproductive isolation (e.g., ref. 21) may represent such equilibria. This provides an alternative explanation for the phenomena of parapatric distributions and the nature of their stability.

The two most influential opponents of the reinforcement model of speciation were Herman Muller and Ernst Mayr. Muller[22] was only prepared to accept that reinforcement was possible when hybrids were

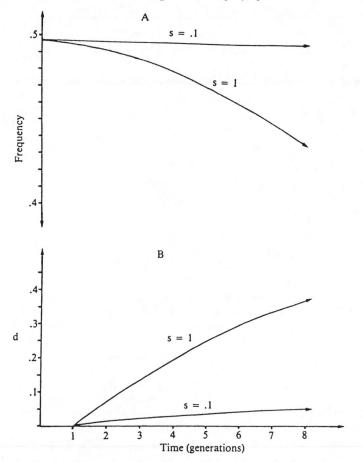

Figure 2. Curves illustrating the effects of changing selective value on the rate of change of *d* (degree of positive assortative mating).

completely disadvantageous. However, Figure 1 has shown that a *k* value appropriate to *D. melanogaster* causes rapid extinction of the rarer population under these conditions. Mayr[23] (p. 551) maintained that, under the conditions of secondary overlap, "there will be strong selection in favour of the acquisition of additional isolating mechanisms to prevent such waste of gametes." With reference to the equation describing the rate of change of *d*, it can be seen that with almost positive assortative mating (i.e., *d* is almost equal to unity) the rate of divergence of *d* under selection is minimal.

In order that their argument be convincing, proponents of the reinforcement model should produce examples of systems of specific-mate recognition which are characterized by *k* values of approximately

0.15 or greater. It has been shown here, however, that in the only case[9] where a reliable estimate was obtainable, the k value found was approximately one order of magnitude less than the minimum k necessary to cause reinforcement.

A number of authors have recognized that lack of convincing cases of "reproductive character displacement" in the literature.[23-25] In view of the strict conditions necessary for the attainment of species status by reinforcement, as outlined in this simulation, perhaps this is not surprising. This simulation suggests that there are substantial problems with the reinforcement model.

CONCLUSION

We have simulated the conditions acting on two populations in secondary contact that have diverged in such a way that hybrids between them are less fit than either parental population. The simulation predicts the effects of dynamic interplay between forces favoring extinction due to heterozygote disadvantage and those favoring the evolution of divergent systems of specific-mate recognition. The results of the "race" between these two competing forces are dependent upon a natural characteristic determining the rate of change of these genetically determined systems of mate recognition. We have termed the constant k a measure of this characteristic. Different values of k influence the system such that a range of outcomes from complete reinforcement to rapid extinction results. A k value of approximately 0.15 or greater results in successful reinforcement. In general, then, if animal or plant species are characterized by genetically determined systems of mate recognition with a k value of this magnitude, the evolution of species by the direct action of natural selection would be possible. One experiment designed to illustrate the divergence in specific-mate recognition systems by natural selection enabled a direct estimate of k for some biological systems. The value calculated from this experiment for *Drosophila melanogaster* was 0.016. This is approximately an order of magnitude smaller than that necessary to cause reinforcement.

We are grateful to many of our colleagues in the Zoology Department for their advice during the course of this study. In particular we would like to thank Leslie Rickett for her early help with the simulation.

REFERENCES

1. Grant, V. 1966. The selective origin of incompatibility barriers in the plant genus *Gilia*. *American Naturalist* 100:1099–1180.

2. MacArthur, R. H. and E. O. Wilson. 1967. *The Theory of Island Biogeography.* Princeton, N.J.: Princeton University Press.

3. Lewontin, R. C. 1974. *The Genetics of the Evolutionary Process.* New York: Columbia University Press.

4. Koopman, K. F. 1950. Natural selection for reproductive isolation between *Drosophila pseudoobscura* and *Drosophila persimilis. Evolution* 4:135–48.

5. Wallace, B. 1950. An experiment on sexual isolation. *Drosophila Information Service* 24:94–96.

6. Knight, G. R., A. Robertson, and C. H. Waddington. 1956. Selection for sexual isolation within a species. *Evolution* 10:14–22.

7. Paterniani, E. 1969. Selection for reproductive isolation between two populations of maize, *Zea mays* L. *Evolution* 23:534–47.

8. Ehrman, L. 1971. Natural selection for the origin of reproductive isolation. *American Naturalist* 105:479–83. Ehrman, L. 1979. Still more on natural selection for the origin of reproductive isolation. *American Naturalist* 113:148–50.

9. Crossley, S. A. 1974. Changes in mating behaviour produced by selection for ethological isolation between ebony and vestigial mutants of *Drosophila melanogaster. Evolution* 28:631–47.

10. Paterson, H.E.H. 1978. More evidence against speciation by reinforcement. *South African Journal of Science* 74:369–71. [This volume, chap. 3.] Paterson, H.E.H. 1980. A comment on "Mate Recognition Systems." *Evolution* 34:330–31. [This volume, chap. 4.]

11. Ayala, F. J. 1975. Genetic differentiation during the speciation process. *Evolutionary Biology* 8:1–78.

12. Harper, A. A., and D. M. Lambert. 1983. The population genetics of reinforcing selection. *Genetica* 62:15–23.

13. Wilson, E. O. 1965. The Challenge from Related Species. Chap. 2 in *The Genetics of Colonizing Species*, ed. H. G. Baker and G. L. Stebbins, 7–27. New York: Academic Press.

14. Bossert, W. H. 1963. Simulation of Character Displacement in Animals. Ph.D. diss., Division of Engineering and Applied Mathematics, Harvard University. Cambridge, Mass.: Harvard University.

15. Key, K.H.L. 1974. Speciation in the Australian Morabine Grasshoppers—Taxonomy and Ecology. In *Genetic Mechanisms of Speciation in Insects*, ed. M.J.D. White, 43–56. Sydney: Australia and New Zealand Book Company.

16. Zaslavsky, V. A. 1966. Isolating mechanisms and its role in the ecology of two allied *Chilocorus* species. *Zoologicheskii zhurnal* 45:203–11.

17. Moran, C., and D. D. Shaw. 1977. Population cytogenetics of the genus *Caledia* (Orthopetra: Acridinae), III. Chromosomal polymorphism, racial parapatry and introgression. *Chromosoma (Berlin)* 63:181–204.

18. Caisse, M., and J. Antonovics. 1978. Evolution in closely adjacent plant populations, IX. Evolution of reproductive isolation in clinal populations. *Heredity* 40:371–84.

19. Crosby, J. L. 1970. The evolution of genetic discontinuity: Computer

models of the selection of barriers to interbreeding between species. *Heredity* 25:253–97.

20. Paterson, H.E.H. 1976. The role of postmating isolation in evolution. Invited Lecture Fifteenth International Congress of Entomology, Washington, Symposium on the Application of Genetics to the Analyses of Species Differences. Unpublished MS [This volume, chap. 2.]

21. Ehrman, L. 1965. Direct observation of sexual isolation between allopatric and between sympatric strains of the different *Drosophila paulistorum* races. *Evolution* 19:459–64.

22. Muller, H. J. 1942. Isolating mechanisms, evolution and temperature. *Biological Symposia* 6:71–125.

23. Mayr, E. 1963. *Animal Species and Evolution.* Cambridge, Mass.: Belknap Press of Harvard University Press.

24. Loftus-Hills, J. J. 1975. The evidence for reproductive character displacement between the toads *Bufo americanus* and *B. woodhousii fowleri. Evolution* 29:368–69.

25. Jackson, J. F. 1973. The phenetics and ecology of a narrow hybrid zone. *Evolution* 27:58–68.

APPENDIX

[p. 311] The following symbols are used below:

N_{AA}, N_{AB} and N_{BB} denote the numbers of the three genotypes. N denotes the combined population number. A symbol such as N' indicates a value for the following generation.

d_A is the probability that an individual of genotype AA will not mate with either an AB or BB individual. d_B is similarly defined, while d is used where $d_A = d_B$.

1. Selection for change in d

The cost of nonassortative mating to an individual of type AA is proportional to the probabilities of:

I. AA meeting BB;	N_{BB}/N	and . . .
accepting BB as a mate;	$(1 - d_A)$	and . . .
being accepted by BB;	$(1 - d_B)$	and . . .
offspring not surviving to reproductive age;	s	
		or . . .
II. AA meeting AB;	N_{AB}/N	and . . .
accepting AB as a mate;	$(1 - d_A)$	and . . .
offspring not surviving to reproductive age;	$1s/2$,	

giving a compound probability:

$$(1/N) \, s \, (1 - d_A) \, [N_{BB} \, (1 - d_B) + N_{AB}] \qquad (1.1)$$

and the increase in d_A in one generation is equal to

$$(1/N) \; ks \; (1 \: - \: d_A) \; [N_{BB} \; (1 \: - \: d_B) \: + \: N_{AB}] \;, \tag{1.2}$$

where k is determined from selection experiments.

Similarly, the increase in d_B is

$$(1/N) \; ks \; (1 \: - \: d_B) \; [N_{AA} \; (1 \: - \: d_B) \: + \: N_{AB}] \;. \tag{1.3}$$

2. Difference equations for the three genotypes

The equations for determining the postselection numbers of the three genotypes in the following generation are:

Letting $D = (d_A \: + \: d_B \: - \: d_A d_B)$,

$$N'_{AA} = (1/N) \; [N^2_{AA} \: + \: N_{AA}N_{AB} \: + \: (1/4)N^2_{AB} \: + \: N_{AA}N_{BB}D] \tag{2.1}$$

$$N'_{AB} = (1/N) \; [N_{AA}N_{AB} \: + \: (1/2)N^2_{AB} \: + \: N_{AB}N_{BB} \: +$$

$$2N_{AA}N_{BB} \; (1 \: - \: D)] \; (1 \: - \: s) \;, \tag{2.2}$$

and

$$N'_{BB} = (1/N) \; [N^2_{BB} \: + \: N_{AB}N_{BB} \: + \: (1/4)N^2_{AB} \: + \: N_{AA}N_{BB}D] \;. \tag{2.3}$$

The "gene" frequencies shown in Figures 1 and 2 are calculated as follows:

$$p = (1/N) \; [N_{AA} \: + \: (1/2)N_{AB}] \tag{2.4}$$

and

$$q = 1 \: - \: p \;. \tag{2.5}$$

3. Calculation of k

Based on the experimental conditions used by Crossley,[9] $N_{AA} = N_{BB}$, $N_{AB} = 0$, and $s = 0$. Therefore, assuming $d_A = d_B = 0$, $d = 0.5k$ by substituting the above values in Equation (1.2). Therefore,

$$k = 2d \;. \tag{3.1}$$

The above is an overestimate since it follows that as d increases so Δd decreases.

Crossley's estimate of random mating was based on the percentage of hybrids obtained; therefore 50 percent hybrids is equivalent to $d = 0$ in her experimental design.

Let H represent the percentage of heterozygotes produced in a cross; then

$$d = -(1/50)H \: + \: 1 \;. \tag{3.2}$$

Taking Crossley's best result, the "dark selection" experiment, and interpolating between $H = 50$ percent and $H = 20$ percent, for $\Delta H = 30$ percent, and, by substitution in Equation (3.2), $d = 0.4$ in fifty generations, which results in an average Δd of 0.008 per generation. Substituting in Equation (3.1) yields $k = 0.016$.

12 The Recognition Concept of Species

Paterson, H.E.H. 1985. The Recognition Concept of Species. In *Species and Speciation*, ed. E. S. Vrba, 21–29. Transvaal Museum Monograph No. 4. Pretoria: Transvaal Museum.

The Symposium on Species and Speciation organized by Elisabeth Vrba at the Transvaal Museum in September 1982 was an important occasion as it brought Niles Eldredge to Pretoria. It also served to bring together most South Africans who are interested in species. It was an exuberant occasion, and it is fortunate that, through Elisabeth Vrba's energy, many of the papers were published in a Transvaal Museum Monograph. Due to the expiry of time, and Elisabeth's policy of allowing the delegates to revise their papers, "The Recognition Concept" reflects the state of the theory in early 1984.

This chapter is important as my first attempt at a broad review of the recognition concept, including an indication of the areas of population biology on which it sheds light. This paper played an important role in making the recognition concept known to a large audience. Unfortunately, it is the only paper of mine which many evolutionists appear to have read.

ABSTRACT

Wallace's view of species has prevailed to this day in a modified form as the so-called biological species concept or isolation concept of species. This concept suffers from serious problems, since it cannot be made to accommodate the only Class I (empirically supported) model of speciation, geographic or allopatric speciation. Problems arise and are outlined with the Class II model of speciation proposed to accommodate the isolation concept: speciation by the reinforcement of isolating mechanisms. All these difficulties fall away if a relational concept of species is abandoned. The recognition concept of species is outlined in some detail for the first time, and the mode of allopatric speciation it determines is discussed. Its wide significance for various fields of population biology is outlined.

INTRODUCTION

> The traditional conception of nature was overthrown by Darwin in 1859, but the fundamental message has never really been understood. The old way of thinking and the new can never be reconciled, but efforts to do so continue to the present day and the results produce a warped and largely erroneous view of the living world. (Ghiselin, 1974)

When we trade in ideas in population biology, species constitute our currency. Their peculiar significance and universality was long ago appreciated by Darwin's mentor, Sir Charles Lyell: "The ordinary naturalist is not sufficiently aware that, when dogmatizing on what species are, he is grappling with the whole question of the organic world and its connection with a time past and with man; that it involves the question of man and his relation to the brutes, of instinct, intelligence and reason, of Creation, transmutation and progressive improvement or development. Each set of geological questions and of ethological and zool.^{L} and botan.^{L} are parts of the great problem which is always assuming a new aspect" (Wilson, 1970:164).

Any view of species must be cast in genetical terms if it is to be useful in understanding the process of evolution, and it must be comprehensive enough to cover most eukaryotes, including man. However, aiming to provide a concept that effectively encompasses the sweep of living forms from unicellular algae through the fungi, plants, and animals to man is indeed an ambitious enterprise. It is, therefore, not really surprising that an entirely satisfactory outcome has not yet been achieved. Like Mayr (1963:426) I believe it is possible that it can be achieved, provided one is able to identify correctly the genetical species' fundamental basis.

In this paper I first consider the broad specifications necessary for any genetical concept of species, then examine the current paradigm and note its shortcomings and inconsistencies. Finally, I offer an alternative view in more detail than previously. This alternative is free from the difficulties that attend the prevailing biological species concept and generates distinct testable predictions, with pervasive implications for all branches of population biology.

With space strictly limited, I here do no more than deal with the main ideas of the two alternative views of genetical species. This is not the place for the detailed documentation of the new recognition concept; this will require book-length treatment, as did the isolation concept in its day (Dobzhansky, 1937; Mayr, 1942). I have been obliged to restrict my consideration of the isolation concept (biological species concept) to the views of the two grand masters, Dobzhansky and Mayr, and have neglected to use the views of other leaders such as Carson and his associates, White, Stebbins, Ayala, and others, despite their significance. Similarly, I have been unable to do more than touch on species and speciation in kingdoms other than Animalia. In particular, special attention will have to be devoted to what happens in plants in a future work. Their relative neglect here should not be taken as implying that I believe them to warrant no closer attention.

THE HEART OF THE GENETICAL SPECIES

The very heart of any genetical species concept, however it is conceived, is vividly revealed by Carson's (1957) apt phrase: "The species as a field for gene recombination." Dobzhansky seems to have had essentially the same idea in his mind when he spoke (1951) of the species as "the most inclusive Mendelian population." Both authors restricted the term to biparental sexual eukaryotes, but otherwise their view seems entirely noncontroversial. Controversy, however, does arise over the way the following two questions should be answered.

1. What sets the limits to the field for gene recombination?
2. How do these limits arise ab initio?

[p. 22] Various genetical species concepts are distinguished by the way they answer these two questions. The concepts to be outlined and examined here are the currently orthodox biological species concept (= isolation concept) and the newer recognition concept of species (Paterson, 1973, 1978 [chap. 3]). My aim is to demonstrate the inconsistencies involved in the former and the advantages of the latter concept.

THE ISOLATION CONCEPT OF SPECIES

It was not until Dobzhansky coined the term "isolating mechanisms" and devoted an entire chapter to them in his classic *Genetics and the Origin of Species* (1937) that the full significance of these adaptations was recognized by evolutionists. (Mayr, 1976:129)

I refer to the prevailing concept of species as the isolation concept because this name explicitly specifies the way the field for gene recombination is supposed to be achieved, namely through the development of the *ad hoc* characters that were referred to by Dobzhansky (1935, 1937), and almost all later authors, as isolating mechanisms.

Thus, for Dobzhansky (1970:357), "species are . . . systems of populations; the gene exchange between these systems is limited or prevented by a reproductive isolating mechanism or perhaps by a combination of several such mechanisms." It will be noticed that, in accordance with this view, a species is defined in terms of its relationship to other species, i.e., it is a *relational* concept, as has long been emphasized by Mayr (1942, 1963, 1970).

Mayr and Provine (1980:34) expressed a similar view: "The major intrinsic attribute characterizing a species is its set of isolating mechanisms that keeps it distinct from other species." These quotations make it evident that both Dobzhansky and Mayr would have answered the first question raised above with: "isolating mechanisms" (Mayr, 1976:520).

If the isolating mechanisms do indeed set the limits of a field for gene recombination, the question arises as to how these characters are acquired by all members of the species. Two obvious circumstances could lead to the fixation of the alleles involved: chance in small, isolated populations and natural selection.

Alleles or chromosomal rearrangements can become fixed by chance even if they are somewhat disadvantageous as heterozygotes. This is clear because pairs of closely related species are known which differ from each other by the fixation of alternative chromosomal arrangements, such as paracentric or pericentric inversions, or by some form of translocation, all of which are more or less disadvantageous in the heterozygous state in most species.

The well-established, and now reasonably understood, phenomenon of pleiotropy is also important in understanding the spread of alleles under selection. In general an allele has more than one phenotypic effect. Some of these may be advantageous, some disadvantageous, and some almost neutral. For an allele to spread by selection, there must be a net selective advantage to it in the heterozygous state. To understand the fixation of an allele under selection, it is essential to understand the nature of its advantage. One would fail to understand how an allele became fixed if, for example, one was aware of only a disadvantageous phenotypic side effect. This type of misjudgment is all too easy to make, since the most conspicuous phenotypic effect may not be the advantageous one, which might well be physiological in character and therefore cryptic. Williams (1966) looked at pleiotropy from an evolutionary viewpoint and restricted the word "function" to apply to the principal advantageous phenotypic effect of an allele. All the remaining consequences, whether advantageous, disadvantageous, or somewhat less advantageous, he named "effects." Thus, in seeking to elucidate the evolution of a species-specific character, we need first to decide whether it was fixed stochastically or by natural selection. If the latter, we need to identify the function it serves. This may be far from obvious, and our judgment may well be colored by our point of view. For example, if we find two sympatric, congeneric species which breed at quite distinct times of the year, an evolutionist might seek a function in terms of *reproductive isolation* and call the character an isolating mechanism. On the other hand, an ecologist might point out that each species breeds when its food is most abundant and might not even notice that it has any influence on the possibility of the members of the two populations hybridizing. Is reproductive isolation the character's *function* or merely an *effect*? I find it doubtful, to put it very mildly, that any "premating isolating mechanism" was ever evolved to serve the function of reproductively isolating the members of one species from those of another. To me, the isolating role seems always to be better accounted for if it is viewed as an effect. Quite certainly the "postmating isolating mechanisms" isolate only incidentally, as was made abundantly clear by Darwin (1859:245 ff.), though a remarkable number of modern population geneticists have failed to see his point. With this argument in mind we can attend to the second question.

Accepting that the field of gene recombination is delimited by the isolating mechanisms imposes restrictions on the answering of the second question: "How do these limits [isolating mechanisms] arise ab initio?" These constraints demand that characters evolved to function as isolating mechanisms must arise in sympatry or parapatry. But this conflicts with the empirical observation that the only Class I (i.e., empirically supported) mode of speciation is speciation in allopatry (Paterson, 1981 [chap. 5]). The difficulties arising from this conflict will be illustrated with quotations from the literature.

The two seminal advocates of the isolation concept, Dobzhansky and Mayr, have both been surprisingly inconsistent in their writings on isolating mechanisms, which makes it difficult to be certain what their concepts actually are. In part this was due to the conflation of the different species concepts (taxonomic, isolation, and recognition) (e.g., Mayr, 1963:89, 95), but other conceptual problems are also involved. Even in his very first theoretical paper on species, Dobzhansky had [p. 23] difficulties. In discussing isolating mechanisms, he wrote (1935:349):

> Although the mechanisms preventing free and unlimited interbreeding of related forms are as yet little understood, it is already clear enough that a large number of very different mechanisms of this kind are functioning in nature. This diversity of the isolating mechanisms is in itself remarkable and difficult to explain. It is unclear how such mechanisms

can be created at all by natural selection, that is what use the organism derives directly from their development. We are almost forced to conjecture that the isolating mechanisms are merely by-products of some other differences between the organisms in question, these latter differences having some adaptive value and consequently being subject to natural selection.

It will be noticed how the acceptance of speciation in allopatry compels authors to invoke pleiotropy as a basis for changing the limits of the field of gene recombination. But of course, isolating mechanisms can then no longer be called "mechanisms," or be regarded as ad hoc, nor can speciation be regarded as adaptive per se. This is the price authors have been reluctant to pay.

The following passage shows that in 1970 Dobzhansky still had fundamental difficulties with this problem (p. 376):

> The question naturally presents itself, what causes bring about the development of reproductive isolating mechanisms? Two hypothetical answers have been proposed. First, reproductive isolation is a by-product of the accumulation of genetic differences between the diverging races. . . . Second, the isolation is built up by natural selection, when and if the gene exchange between the diverging populations generates recombination products of low fitness . . . These two hypotheses are not mutually exclusive. Needless disputes have arisen because they were mistakenly treated as alternative.

In practice the problem cannot so readily be disposed of. The assertion that these two hypotheses are not alternatives seems to me to be impossible to sustain; the first does not invoke any direct selection for "reproductive isolation," while the second does. This is a very fundamental difference between them, as each has very different consequences for evolutionary theory.

In 1950 Dobzhansky wrote (see also Dobzhansky, 1976:104): "The integration of individuals into Mendelian populations, into sexual supraorganisms if you will, is an evolutionary adaptation. The further integration of elementary Mendelian populations into populations of higher orders, such as races and species, is likewise adaptive. It can be shown that formation of races and species has become necessary owing to the great diversity of environments found on our planet." This statement provides us with an insight on how Dobzhansky conceived the role of species, why he required species to be the direct products of natural selection (see Mayr, 1949:284, for a comparable statement). In the following passage (Dobzhansky, 1940), his bias in favor of the selection hypothesis clearly shows through: "Another difficulty is with species isolated on oceanic islands or in similar situations. If the geographical barriers between races are secure enough, the precondition for the development of physiological [= reproductive] isolation is absent. Precisely how serious is this difficulty is not clear." In the third edition of his book (Dobzhansky, 1951:208) we find: "Isolating mechanisms encountered in nature appear to be *ad hoc* contrivances which prevent the exchange of genes between nascent species, rather than incongruities originating in accidental changes in the gene functions." A clear reluctance to rely on pleiotropy in isolation is evident.

Inevitably, Mayr (1963:548; see also p. 109) was plagued by the same problem: "They are *ad hoc* mechanisms. It is therefore somewhat difficult to comprehend how isolating mechanisms can evolve in isolated populations." Mayr's problem was particularly acute since he is an ardent advocate of geographical or allopatric speciation. I have recently (1981:113 [chap. 5]) emphasized that there is much observational evidence for the origin of new species in total isolation (Mayr, 1942, 1963) and that there is no such support for any other model of speciation. Dobzhansky's favored model of speciation through the reinforcement of isolating mechanisms under natural selection is one among many for which there is no convincing evidence. Even after some forty years of active research, the support for speciation by reinforcement today is still both meager and equivocal (Loftus-Hills, 1975; Paterson, 1978 [chap. 3], 1981 [chap. 5], 1982a [chap. 8]).

These few difficulties alone seem sufficient to falsify the isolation concept of species, and yet they constitute but some of its problems.

The isolation concept was supported by Wallace (1889) but not by Darwin. Later it was further elaborated by Gulick (1905), in remarkably modern terms, who appears to have anticipated Dobzhansky to a considerable extent. Both Robson (1928) and Fisher (1930) supported the isolation concept. Doubtless, however, the concept's wide support stems from Dobzhansky's effective advocacy (1937, 1940, 1970).

As already stated, Dobzhansky was aware of problems and inconsistencies with the concept in 1935, and yet he adhered to it consistently throughout his life. It is important to ask why he should have done so. In a wider context, the very general, unquestioning support the isolation concept has received is suspicious. To me it suggests that this favor might stem from deep-seated biases inherent in our Western cultural background (cf. Masters et al., 1984; Paterson, 1982b [chap. 9]). One obvious source of bias is the creation stories in the book of Genesis. That this has been, and indeed is, an important source of prejudice was made clear by Darwin at the beginning of his chapter on hybridism (Darwin, 1859:245): "The view generally entertained by naturalists is that species . . . have been specially endowed with the quality of sterility in order to prevent the confusion of all organic forms."

Support for this contention can be gleaned from Ellegard (1958), who records the opinions of nineteenth-century general readers and scientists alike: "God forbids hybrids to breed," "the sterility of true hybrids affords another evidence of the jealousy with which the Creator regards all attempts to introduce confusion into His perfect plan," etc. From this widely held viewpoint, [p. 24] a species is seen as a group of organisms descended from the same originally created pair.

Another, even older, source of bias in favor of the isolation concept stems from prehistoric attitudes to "purity of stock" and "purity of line," engendered by the practices of ancient plant and animal breeders. The introduction of managed breeding programs undoubtedly revolutionized man's whole way of life and, one would think, influenced his language by enriching its vocabulary and imagery. In English, notice how approbative are words such as "pure," "purebred," and "thoroughbred," and how pejorative are those like "mongrel," "bastard," "halfbreed," and "hybrid." Such cultural biases, which act subtly, almost subliminally, through the vocabulary and imagery of languages, might well predispose the unwary to favor ideas like that of "isolating mechanisms" with the role of "protecting the integrity of species." When Dobzhansky introduced the isolation concept in 1937 it was accepted almost without resistance. This acceptance is in sharp contrast to the usual opposition that greets new ideas (cf. Kuhn, 1970) and could well have been due to these cultural, predisposing factors.

Thus, problems of function, logic, and consistency beset the isolation concept.

THE RECOGNITION CONCEPT OF SPECIES

> It is no use for a species to have individuals of two sexes, supplied with reproductive apparatus mutually adapted for their propagation, if the individuals are not endowed with suitable genetically determined impulses. The potential parents must also have instincts, prompting them to desire one another and to provide for the resulting young such care as they need for their survival. (Darlington, 1964:335)

It is rather obvious that the complex act of fertilization, which is an essential part of sex, will not occur fortuitously, as is recognized by Darlington in the quotation above. The physiologist J. Z. Young (1975) provided an independent view of the same situation:

> To ensure that reproduction occurs at times of the year when young are likely to survive, the slowly operating chemical signalling of the endocrine

system is used. . . . Control through the nervous system is important, however, for the performance of reproductive actions in relation to other members of the species, and in all mammals there are communication signs by which individuals recognize others of the same or opposite sex and elicit appropriate mating reactions from them. . . . Further, there are *secondary sexual characters* providing signals ensuring that the sexes meet and pair.

Among evolutionists, recognition of the existence of a fertilization system in all biparental eukaryotes is general, though its full significance is seldom emphasized. Thus, both Dobzhansky (1951:184) and Mayr (1963:95) were aware of its existence. What has not been recognized and emphasized is that the form of the fertilization system accounts for the delineation of "fields for gene recombination" in a simple, rational, and yet comprehensively adequate way. Mayr (1963:20) in fact rejects this approach explicitly in favor of a relational concept: "Species are more unequivocally defined by their relation to non-conspecific populations ('isolation') than by the relation of conspecific individuals to each other. The decisive criterion is not the fertility of individuals but the reproductive isolation of populations."

True sex is found only in biparental, eukaryotic organisms. Despite considerable variation in detail, sexual cycles always involve the two complex processes of meiosis and fertilization (Lewis and John, 1964). Successful fertilization in even the simplest of unicellular eukaryotes requires the assistance of a series of adaptations which constitute what might be called the fertilization system of the organism (Fig. 1).

Each fertilization system comprises a number of components, each adapted to fulfill a subsidiary proximate function, which contributes to the ultimate function of bringing about fertilization while the organism occupies its normal habitat.

Fertilization systems are also adapted to the circumstances imposed by the organism's way of life (i.e., whether sessile, motile, nocturnal, diurnal, etc.). Thus, motile organisms have a special requirement for a subset of characters of the fertilization system that serve the function of bringing motile mating partners together as a necessary preliminary to the ultimate achievement of fertilization and syngamy.

This subset of adaptations, which are involved in signaling between mating partners or their cells, constitutes the specific-mate recognition system or SMRS (Fig. 2) (Paterson, 1978 [chap. 3], 1980 [chap. 4]). I have chosen to use "specific-mate" instead of simply "mate," because it is important to distinguish between the process with which I am concerned and the quite different kind of mate recognition, "individual mate recognition." I also avoid the common term "species recognition" because the process with which I am concerned most certainly does not involve "species recognition," a strictly philosophical

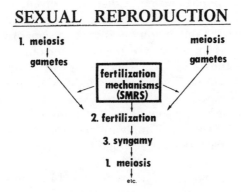

Figure 1. Sexual cycle in a biparental eukaryote.

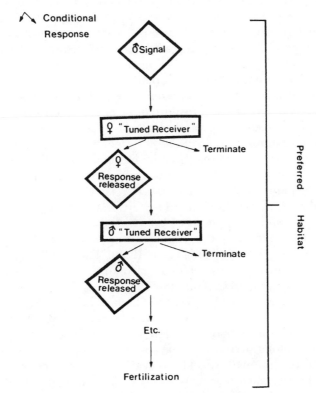

Figure 2. Simplified diagram illustrating the form of the specific-mate recognition system (SMRS) as a coadapted signal-response reaction chain. A more realistic representation was given by Morris (1970:338).

process, but the recognition of an appropriate mating partner. ("Appropriate" here implies no more than an individual of opposite sex drawn from the same "field for gene recombination.") [p. 25]

The response of one mating partner to a signal from the other is here regarded as an act of *recognition* (Paterson, 1982a [chap. 8]). Recognition is thus a specific response by one partner to a specific signal from the other. This process might be compared to the recognition of a specific antigen by its specific antibody, and hybridization involves a process having something in common with a cross-reaction. I strongly emphasize that I imply no act of judgment and no act of choice on the part of the responding partner (Paterson, 1982a [chap. 8]).

As a working hypothesis I have assumed that all steps in a fertilization system have evolved to serve a particular role. In other words, I assume that all components are adaptations, or, at least, were adaptations, in the strict sense of Williams (1966). It is possible to sustain such a view if one examines the well-analyzed "courtships" of the three-spined and ten-spined sticklebacks (Tinbergen, 1951; Morris, 1970), though, admittedly, the case is less diagrammatically clear in species which have not yet been as closely analyzed (e.g., the ducks).

In sessile organisms the SMRS assumes a much less prominent role; positive assortative mating is determined in these organisms by other adaptations of the fertilization system. The sex cells of sessile biparental eukaryotes are transported by various agencies, including wind, water, insects, birds, and mammals (e.g., bats). In sessile organisms, as in motile, the fertilization system is closely adapted to the conditions of the organism's way of life and normal habitat. This becomes strikingly evident when one compares, for example, the fertilization systems of two very distantly related organisms from the intertidal zone of a rocky shore. The fertilization system of an oyster (or a mussel) and of a species of brown alga are much more similar than are the systems of the oyster and its motile predatory fellow mollusc, *Sepia officinalis* (Tinbergen, 1939). The fertilization system of the squid, in turn, resembles much more closely the systems of some fish (Packard, 1972; Ghiselin, 1974).

Among angiosperms the SMRS is limited to interactions such as between the pollen and the stigma (Clarke and Knox, 1978), and in sessile animals it may be restricted to the recognition of the sperm by the ovum. Among orchids, several species, even of different genera, may share a common SMRS. This is evident from the observation that fertile experimental crosses can be made between them. In these species the "fields for gene recombination" are determined by the fertilization mechanisms other than those of the SMRS. In orchids they usually involve the attraction and exploitation of a particular insect species as a vector of pollen (e.g., the Western Australian slipper orchid, *Cryptostylus ovata*, Erickson, 1951). Darwin, in the introductory sentences of his book (1862) on the pollination of orchids by insects, expresses a view in perfect agreement with the recognition concept: "The object of the following work is to show that the contrivances by which the orchids are fertilized are varied and almost as perfect as any of the most beautiful adaptations in the animal kingdom; and, secondly, to show that these contrivances have for their main object the fertilization of the flowers with pollen brought by insects from a distant plant."

Taking such facts as these into account, it is clear that the limits to a "field for gene recombination" can be set by the shared fertilization system of members of a population of organisms alone. It is the fertilization system which leads to the positive assortative mating that delimits the field.

We can, therefore, regard as a species *that most inclusive population of individual biparental organisms which share a common fertilization system.*

I now turn to the second question: How does a new fertilization system, and hence a new species, arise ab initio?

In terms of fitness, a mate is a crucial resource. Mate acquisition is, thus, as important an adaptive process as is avoidance of predators, the ability to find a refuge, or the acquisition of food. As with these other adaptive characters, the characters of the fertilization system are adapted to the circumstances impinging on the organism in its normal habitat. While members of a species are occupying their normal habitat, the characters of the fertilization system, as well as their other adaptive characters, are all maintained under stabilizing selection. This means that while biparental eukaryotes remain in their normal habitats, their adaptive characters remain largely buffered from directional change.

From this insight, it is evident that for speciation to occur the buffering of adaptive characters must be overcome, [p. 26] so that a new constellation of adaptive character-states can evolve. This is most likely to happen if a small population (Paterson and Macnamara, 1984:315 [chap. 10]) of conspecific individuals becomes displaced into, and restricted to, a new habitat. The isolation envisaged here is allopatric isolation induced by extrinsic circumstances, such as a flock of birds being blown out to sea onto an island on which their normal habitat is not to be found. Many other scenarios might be invoked according to particular circumstances. Any adaptive characters, including those of the fertilization system, which are now less well adapted, will become subject to directional selection. Eventually, when the character-states of the population members have shifted under selection, etc., to become appropriate and effective under the new conditions, they will return once more to the control of stabilizing selection. At this point we say that the individual organisms of the population are adapted to the new habitat. Speciation will also have occurred if the new fertilization system has become sufficiently different from that of the members of the parent population, for then the new fertilization system will delimit a new field for gene recombination. At present the time needed for such speciation to occur is essentially unknown, but it may be relatively short (Greenwood, 1965).

The adaptation of all the adaptive characters to the conditions of the new habitat also determines a new "niche" for the daughter species, and the stabilizing selection within the normal habitat stabilizes these ecological characteristics as well. It is clear that speciation and the acquisition of a new niche are part of the same process of adaptation to a new habitat. There is no special process of cladogenesis; speciation is an incidental *effect* resulting from the adaptation of the characters of the fertilization system, among others, to a new habitat, or way of life.

Although this view of speciation must be classified with Mayr's model of geographical speciation (Mayr, 1963) it should be appreciated that they are not identical. Because Mayr conceives species in terms of reproductive isolation, he is obliged to invoke the pleiotropic modification of "isolating mechanisms" in allopatry to account for geographic speciation. No such problem faces the advocates of the new species concept. Since the characters of the fertilization system are clearly adaptive in relation to the species' normal habitat, speciation results from their adaptation to the conditions in the new habitat (see below).

Despite a major role being assigned to natural selection in the model of speciation discussed above, I do not count myself in the ranks of Kimura's "naive panselectionists." I do accept an important role for stochastic processes. This view can scarcely be avoided if one notices that closely related species may differ in possessing fixed alternative chromosomal arrangements (see above).

The correlation of fixed chromosome rearrangements with speciation cannot be regarded as evidence for a causal connection between the two events as White (1978) has argued (cf. Carson et al., 1967; Paterson, 1981 [chap. 5]).

A small population isolated in a new habitat will more often than not become extinct. However, should it survive over a number of generations it is likely to remain small because its members are ill adapted and are under directional selection. The population is likely to increase only after its members have adapted to the new conditions.

It is unlikely that the characters of the SMRS can adapt quickly to new conditions. This is because the coadaptation that exists between the mating partners with respect to signals and receivers constitutes an effective buffer to change (Paterson, 1978 [chap. 3]). The signal and receiver can only change in small steps, with coadaptation being reestablished after each step. The details of this process deserve a more comprehensive treatment than can be offered here.

DISCUSSION

In reviewing the consequences of the recognition concept, the first thing to consider is the proposition that the isolation and recognition concepts of species are merely "the two sides of the same coin."

According to the proponents of the isolation concept, species are determined by the functioning of the isolating mechanisms; that is, species are defined relationally in terms of reproductive isolation. Leading workers have regarded species explicitly, though inconsistently, as "adaptive devices."

In terms of the recognition concept, on the other hand, species are determined by the functioning of fertilization mechanisms; species are not defined relationally but independently, since, of necessity, all sexual organisms must possess an effective fertilization system. Species are seen as incidental consequences of adaptive evolution.

Applying the two concepts in turn to a newly arisen autotetraploid (see also Paterson, 1981 [chap. 5]) in a population of an outbreeding diploid angiosperm will serve to show that they are in fact quite distinct and lead to quite different expectations. The situation described is generally accepted as a case of "instantaneous speciation" by advocates of the isolation concept (e.g., by Mayr, 1963:439, 448), because the tetraploid individuals are "reproductively isolated" from the parental diploid population. In genetical terms, however, the gene pool of the tetraploids is merely a subset of the gene pool of the diploids. Accordingly, "mating" is likely to occur at random. If this is considered in the light of population genetic theory, it will be seen that we are here dealing with an example of heterozygote disadvantage. The deterministic algebra makes it abundantly clear that, under the conditions outlined above, the tetraploid population would be rapidly eliminated. Inability to coexist is scarcely a criterion for a species in population genetics.

On the other hand, if the recognition concept is applied to this situation, the conclusion is inevitably reached that the two populations are conspecific because they share a common fertilization system.

Natural selection, as shown above, thus eliminates the cause of the heterozygote disadvantage within the single species.

Besides revealing the distinctness of the two concepts, [p. 27] this scenario underlines a point which is not generally appreciated by evolutionists. This is that two populations that share a common SMRS cannot coexist if they are technically "reproductively isolated" by "sterility" alone (Paterson, 1968). In such a case, with complete interpopulation sterility ($s = 1$), the less common population will rapidly disappear (Paterson, 1978 [chap. 3]); but if the heterozygote disadvantage is less complete (i.e., if the coefficient of selection, s, is less than 1 but greater than 0), natural selection will act to eliminate the allele or chromosome rearrangement that is causing the hybrid disadvantage. Under these conditions, the "interpopulation sterility" is thus an intraspecific phenomenon, comparable to the phenomenon of "self-incompatibility" in plants (Clarke and Knox, 1978).

An example from animals is the well-known *Drosophila paulistorum* (Dobzhansky, 1970). Subpopulations are often semi-intersterile, despite sharing a common SMRS (in some cases they mate at random). The infertility is due to the interaction of the host genome with that of symbiotic mycoplasms in the gonads. In accordance with the recognition concept these subpopulations are conspecific, but Dobzhansky interpreted them as "semispecies." A comparable situation is found in the mosquito species *Culex molestus* Forskål of the *Culex pipiens* complex. The symbionts in this case are rickettsias (Irving-Bell, 1983; Miles, 1977). Some of these cases, at least, are laboratory artifacts arising from "bottlenecks" in the colonies. Studies with fresh material from the field should be done in both complexes. Theoretical models of "sympatric speciation" which depend on one or other form of "sterility" (heterozygote disadvantage) are fairly widely accepted by evolutionists. But, again, in many cases they do not even constitute speciation according to the recognition concept because both populations in these models share a common fertilization system. Summarizing this point: under the recognition concept, all phenomena covered by the category "postmating isolating mechanisms" (Mayr, 1963) are incidental to delineating species, since they have nothing to do with bringing about fertilization.

It has been shown that the recognition concept is not simply the "obverse side of the same coin" as the isolation concept and some of the logical and interpretive difficulties of the isolation concept have been enumerated. How well does the recognition concept accommodate currently debated ideas in population biology?

Reporting on the recent Chicago Macroevolution Conference, Maynard Smith (1981) made the following comment on the pattern of evolutionary change known as the punctuated equilibria model (Eldredge and Gould, 1972): "As a geneticist I was left with the impression that the 'sudden' appearance of a new species called for no new explanations, but that stasis may do." In fact the recognition concept, and the mode of allopatric speciation it specifies, account for all aspects of the punctuated equilibria model very satisfactorily. The equilibrium is well explained by the stabilizing selection due to the coadaptation of the SMRS between the mating partners in their normal habitat, as has been explained. The "sudden appearance" of new species fits a speciation model in which small, isolated populations develop into new species and grow in population numbers only after stabilizing selection is again in effect on the new species. The best illustration of these events that I know of from the fossil record is Coope's (1979) documentation of Cenozoic beetles moving with their habitat. The isolation concept does not lead one to expect equilibrium following a speciation event, as Maynard Smith's comment implies.

A point of special interest is that the recognition concept leads one to expect that in those species where some of the SMRS signals are preserved (e.g., horns or antlers), the fossil record is likely to illustrate punctuated equilibria with least ambiguity (Paterson, 1982a [chap. 8]; Vrba, 1984).

The stability of adaptive characters expected under the recognition concept is relevant to the assessment of Van Valen's (1973) Red Queen hypothesis (see Paterson, 1982a [chap. 8]). The recognition concept and Coope's observations on fossil beetle populations suggest that Van Valen's expectation that natural selection acts mainly "to enable the organisms to maintain their state of adaptation rather than to improve it" is not generally valid and may only apply under special circumstances. If Van Valen's hypothesis was, in fact, realistically conceived, the punctuated equilibria pattern would not be expected as an outcome.

It is not always appreciated that the nature of species and the way speciation occurs have a fundamental role in any attempt at understanding the structure of communities and ecosystems. This is because most ecologists and population biologists attribute the fundamental causal role of patterns in nature to interspecific competition. However, in recent years a number of workers have come to question the critical role attributed to interspecific competition (e.g., Andrewartha and Birch, 1954; Wiens, 1977; Connell, 1980; Walter, Hulley and Craig, 1984). The field is too large to review in detail here, but it should be noticed that no emphasis was placed on interspecific competition as an environmental factor when considering how speciation occurs. Generally speaking, I agree that interspecific competition lacks the credentials of a "motor" for evolutionary change. After much study there are still no really compelling examples of character displacement (Grine, 1981). While adaptation to a habitat and the coadaptation of the male and female parts of the SMRS are constraints, interspecific competition exerts only intermittent selective pressure on members of a species. Therefore, the evident stability of form in many species in space and time (see discussion above) with respect to adaptive characters does not favor interspecific competition as a selective force to account for "niche" characteristics.

If speciation occurs as postulated above under the recognition concept, then we obtain a radically different picture of how the structures of communities and ecosystems emerge. The normal habitat of a species is likely to approximate closely to that in which it arose. Similarly, the adaptive characters that were fixed in the original small isolate were subsequently stabilized in a [p. 28] number of ways, including stabilizing selection within the normal habitat of the species (cf. Coope, 1979). The increase in numbers of the original small population which speciated also constitutes a buffer to change. Accordingly, members of sympatric species do not come to "partition resources" in an ad hoc manner. Character displacement is not expected according to this model, and it is not surprising that clearcut cases are virtually nonexistent. Species characteristics (fertilization system, niche characteristics, etc.) are expected to be largely stabilized throughout the range of the species. These expectations contrast strongly with those generated by interspecific competition theory, as Walter, Hulley, and Craig (1984) have emphasized.

Although many important consequences could be added to the few examined in this discussion, only one other will be emphasized: the bearing the recognition concept has on how we look at man.

The isolation concept has a very definite teleological flavor to it (Paterson, 1982b [chap. 9]). Species are seen as "adaptive devices," and speciation is considered a process of adaptation by the leaders in the field. Isolating mechanisms are said to be *ad hoc* devices evolved to protect the integrity of a species. Such statements are teleological, or imply the involvement of group selection, or both.

In contrast, the recognition concept is free of teleological commitment and invokes only individual selection, as with Darwin's view (1859). Darwin conceived species in terms of structure, just as taxonomists did. He saw no essential difference between species and varieties. He believed that both arose in the same way, under natural selection. Inadequate though his view of species was, Darwin's view of speciation was detailed enough for us to see that he accepted that species arise as incidental consequences of adaptation. This was perhaps the most revolutionary conclusion in a revolutionary text, yet it is seldom stressed or even appreciated except by philosophers. Here it is emphasized by Hull (1973:55): "Galileo and Newton replaced one physical theory with another, but they left the teleological world-picture intact. The findings of the paleontologists and geologists had necessitated a reinterpretation of Genesis. But it was Darwin who finally forced scientists to realize just how trivial teleology had become in their hands. The change in scientific thought marked by the appearance of the *Origin of Species* was so fundamental that it certainly deserves the title conceptual revolution." In the same vein, Thomas Kuhn (1970:172) said, "For many men the abolition of that teleological kind of evolution was the most significant and least palatable of Darwin's suggestions. The *Origin of Species* recognized no goal set either by God or nature."

Thus, in sharp contrast to the isolation concept, the recognition concept is in complete accord with the revolutionary view of Darwin as far as it went. Moreover, the recognition concept emphasizes the incidental nature of speciation and expresses it in genetical terms, besides providing a genetical concept of species. Applied to man, the concept is equally appropriate.

It was pointed out above that any comprehensive view of species must be able to deal satisfactorily with the question: What sets the limits to the field of gene recombination in man? In applying the isolation concept to man, a number of problems arise. Authors never seem to identify the isolating mechanisms that are supposed to delimit the field. Perhaps there are none, in accordance with this statement by Mayr (1963:109), "Where no other closely related species occur, all courtship signals [= 'ethological isolating mechanisms'] can 'afford' to be general, nonspecific, and variable." However, I find Mayr's assertion to be poorly supported empirically (Paterson, 1978 [chap. 3]). I believe that Mayr wrote these words with a case such as the mallard and pintail ducks and some of their insular congeners in mind (Lack, 1974). It is hypothesized that the mallard and pintail are markedly sexually dimorphic in the Holarctic region to isolate them reproductively from their many close congeners with which they are sympatric. On a number of islands such as Hawaii, Kerguellan, South Georgia, and the Falkland Islands monomorphic relatives exist which are allopatric from all close congeneric species. It is presumed that they can here "afford" to relax their "isolating mechanisms." In 1978 I pointed out that, in Africa, Australia, and North America, a number of monomorphic relatives of the mallard and pintail coexist without requiring the striking sexual dimorphism of their close relatives of the Holarctic, which weakens the case considerably (see also Weller,

1980:68). It is further weakened when one notices that species with no close relatives on earth (the ostrich, the hamerkop [*Scopus umbretta*], the Madagascan partridge [*Margaroperdix madagascariensis*], etc.) nevertheless possess complex courtships. Thus, I believe the case for applying Mayr's statement to man is not convincing.

No such problem occurs in applying the recognition concept to *Homo sapiens*. This is not the place to analyze the SMRS of man in detail, and so I shall merely provide an independent and appealing lay statement of its existence by the psychologist Abraham Maslow (1970:195): "It is not the welfare of the species, or the task of reproduction, or the future development of mankind that attracts people to each other . . . it is basically an enjoyment and a delight, which is another thing altogether."

The isolation concept in its complete form was presented for testing in 1940. Today it is still logically unsatisfactory, and in virtually all accounts it is intricately conflated with either or both the recognition concept and a morphological concept of species. Identifying the recognition concept as fundamentally different from the isolation concept enables us to design stringent tests in attempts at refuting the one or the other. This is definite scientific progress. My earnest hope is that it will reduce the prevailing fog and induce new life into that very significant biological discipline once called by Ernst Mayr the "science of species."

ACKNOWLEDGMENTS

This paper is dedicated to Ernst Mayr, my first guide, for his eightieth year. It is a contribution from the Animal Communications Research Group, which is funded by the Council Research Committee of the University of the Witwatersrand. [p. 29]

For critical comments and guidance, my deep gratitude is due to Michael Anderson, Neil Caithness, Marc Centner, Anthony Gordon, Pat Hulley, Shane McEvey, Judith Masters, Andy Potts, Rob Toms, and Elisabeth Vrba, and a referee. I also thank Adele Katz for her efficient secretarial assistance.

REFERENCES

Andrewartha, H. G., and L. C. Birch. 1954. *The Distribution and Abundance of Animals*. Chicago: University of Chicago Press.
Carson, H. L. 1957. The Species as a Field for Gene Recombination. In *The Species Problem*, ed. E. Mayr, 23–38. Publication No. 50. Washington, D.C.: American Association for the Advancement of Science.

Carson, H. L., F. E. Clayton, and H. D. Stalker. 1967. Karyotypic stability and speciation in Hawaiian *Drosophila*. *Proceedings of the National Academy of Science, USA* 57:1280–85.

Clarke, A. E., and R. B. Knox. 1978. Cell recognition in flowering plants. *Quarterly Review of Biology* 53:3–28.

Connell, J. H. 1980. Diversity and the coevolution of competitors, or the ghost of competition past. *Oikos* 35:131–38.

Coope, G. R. 1979. Late Cenozoic fossil Coleoptera: Evolution, biogeography, and ecology. *Annual Review of Ecology and Systematics* 10:247–67.

Darlington, C. D. 1964. *Genetics and Man*. London: Allen and Unwin.

Darwin, C. 1859. *On the Origin of Species by Means of Natural Selection; or, the Preservation of Favoured Races in the Struggle for Life*. London: John Murray.

———. 1862. *The Various Contrivances by Which Orchids Are Fertilized by Insects*. London: John Murray.

Dobzhansky, T. 1935. A critique of the species concept in biology. *Philosophy of Science* 2:344–55.

———. 1937. *Genetics and the Origin of Species*. 1st ed. New York: Columbia University Press.

———. 1940. Speciation as a stage in evolutionary divergence. *American Naturalist* 74:312–21.

———. 1950. Mendelian populations and their evolution. *American Naturalist* 84:401–18.

———. 1951. *Genetics and the Origin of Species*. 3d ed. New York: Columbia University Press.

———. 1970. *Genetics of the Evolutionary Process*. New York: Columbia University Press.

———. 1976. Organismic and Molecular Aspects of Species Formation. In *Molecular Evolution*, ed. F. J. Ayala, 95–105. Sunderland, Mass.: Sinauer Associates.

Eldredge, N., and S. J. Gould. 1972. Punctuated Equilibria: An Alternative to Phyletic Gradualism. Chap. 5 in *Models in Paleobiology*, ed. T.J.M. Schopf, 82–115. San Francisco: Freeman, Cooper and Company.

Ellegard, A. 1958. *Darwin and the General Reader*. Gothenburg Studies in English. Vol. 8. Göteborg, Sweden: University of Göteborg.

Erickson, R. 1951. *Orchids of the West*. Perth, Western Australia: Paterson Brokensha.

Fisher, R. A. 1930. *The Genetical Theory of Natural Selection*. 1st ed. Oxford: Oxford University Press.

Ghiselin, M. T. 1974. *The Economy of Nature and the Evolution of Sex*. Berkeley: University of California Press.

Greenwood, P. H. 1965. The cichlid fishes of Lake Nabugabo, Uganda. *Bulletin of the British Museum (Natural History) (Zoology)* 12:315–57.

Grine, F. E. 1981. Trophic differences between "Gracile" and "Robust" Australopithecines: A scanning electron microscope analysis of occlusal events. *South African Journal of Science* 77:203–30.

Gulick, J. T. 1905. *Evolution, Racial and Habitual*. Publication No. 25. Washington, D.C.: Carnegie Institute of Washington.

Hull, D. 1973. *Darwin and His Critics.* Cambridge, Mass.: Harvard University Press.

Irving-Bell, R. J. 1983. Cytoplasmic incompatibility within and between *Culex molestus* and *Cx. quinquefasciatus* (Diptera: Culicidae). *Journal of Medical Entomology* 20:44–48.

Kuhn, T. 1970. *The Structure of Scientific Revolutions.* 2d ed. Chicago: University of Chicago Press.

Lack, D. 1974. *Evolution Illustrated by Waterfowl.* Oxford: Blackwell Scientific Publications.

Lewis, K. R., and B. John. 1964. *The Matter of Mendelian Heredity.* London: J. & A. Churchill.

Loftus-Hills, J. J. 1975. The evidence for reproductive character displacement between the toads *Bufo americanus* and *B. woodhousii fowleri. Evolution* 29:368–69.

Maslow, A. 1970. *Motivation and Personality.* New York: Harper and Row.

Masters, J., D. M. Lambert, and H.E.H. Paterson. 1984. Scientific prejudice, reproductive isolation, and apartheid. *Perspectives in Biology and Medicine* 28:107–16.

Maynard Smith, J. 1981. Macroevolution. *Nature* 289:13–14.

Mayr, E. 1942. *Systematics and the Origin of Species.* New York: Columbia University Press.

———. 1949. Speciation and Systematics. In *Genetics, Paleontology, and Evolution,* ed. G. L. Jepsen, E. Mayr, and G. G. Simpson, 281–98. Princeton, N.J.: Princeton University Press.

———. 1963. *Animal Species and Evolution.* Cambridge, Mass.: Belknap Press of Harvard University Press.

———. 1970. *Populations, Species and Evolution.* Cambridge, Mass.: Belknap Press of Harvard University Press.

——— ed. 1976. *Evolution and the Diversity of Life.* Cambridge, Mass.: Belknap Press of Harvard University Press.

Mayr, E., and W. B. Provine. 1980. *The Evolutionary Synthesis.* Cambridge, Mass.: Belknap Press of Harvard University Press.

Miles, S. J. 1977. Laboratory evidence for mate recognition behaviour in a member of the *Culex pipiens* complex (Diptera: Culicidae). *Australian Journal of Zoology* 25:491–98.

Morris, D. 1970. *Patterns of Reproductive Behaviour.* London: Jonathan Cape.

Packard, A. 1972. Cephalopods and fish: The limits of convergence. *Biological Reviews* 47:241–307.

Paterson, H.E.H. 1968. Evolutionary and Population Genetical Studies of Certain Diptera. Ph.D. diss., University of the Witwatersrand, Johannesburg.

———. 1973. Animal species studies. Pp. 31–36 in Animal and plant speciation studies in Western Australia, by H.E.H. Paterson and S. H. James. *Journal of the Royal Society of Western Australia* 56:31–43. [See this volume, Author's Preface.]

———. 1978. More evidence against speciation by reinforcement. *South African Journal of Science* 74:369–71. [This volume, chap. 3.]

————. 1980. A comment on "Mate Recognition Systems." *Evolution* 34:330–31. [This volume, chap. 4.]

————. 1981. The continuing search for the unknown and unknowable: A critique of contemporary ideas on speciation. *South African Journal of Science* 77:113–19. [This volume, chap. 5.]

————. 1982a. Perspective on speciation by reinforcement. *South African Journal of Science* 78:53–57. [This volume, chap. 8.]

————. 1982b. Darwin and the origin of species. *South African Journal of Science* 78:272–75. [This volume, chap. 9.]

Paterson, H.E.H., and M. Macnamara. 1984. The recognition concept of species (Michael Macnamara interviews H.E.H. Paterson). *South African Journal of Science* 80:312–18. [This volume, chap. 10.]

Robson, G. C. 1928. *The Species Problem*. Edinburgh: Oliver and Boyd.

Tinbergen, L. 1939. Zur Fortpflanzungsethologie von *Sepia officinalis* L. *Archives Neerlandaises de Zoologie* 3:323–64.

Tinbergen, N. 1951. *The Study of Instinct*. Oxford: Oxford University Press.

Van Valen, L. 1973. A new evolutionary law. *Evolutionary Theory* 1:1–30.

Vrba, E. S. 1984. Evolutionary Pattern and Process in the Sister-Group Alcelaphini-Aepycerotini (Mammalia: Bovidae). In *Living Fossils*, ed. N. Eldredge and S. M. Stanley, 62–79. New York: Springer-Verlag.

Wallace, A. R. 1889. *Darwinism*. London: Macmillan.

Walter, G. H., P. E. Hulley, and A.J.F.K. Craig. 1984. Speciation, adaptation and interspecific competition. *Oikos* 43:246–48.

Weller, M. W. 1980. *The Island Waterfowl*. Ames: Iowa State University Press.

White, M.J.D. 1978. *Modes of Speciation*. San Francisco: W. H. Freeman.

Wiens, J. A. 1977. On competition and variable environments. *American Scientist* 65:590–97.

Williams, G. C. 1966. *Adaptation and Natural Selection*. Princeton, N.J.: Princeton University Press.

Wilson, L. G. 1970. *Sir Charles Lyell's Scientific Journals on the Species Question*. New Haven, Conn.: Yale University Press.

Young, J. Z. 1975. *The Life of Mammals*. Oxford: Oxford University Press.

13 Environment and Species

Paterson, H.E.H. 1986. Environment and species. *South African Journal of Science* 82:62–65.

I was pleased to have the opportunity to address an important international meeting of paleontologists, paleoclimatologists, and anthropologists on the significance of the recognition concept for the study of past climates and for interpreting pattern in the fossil record in terms of speciation. This paper extends earlier remarks about the critical role of environment in relation to species and speciation.

It is appropriate to consider the nature of species in this workshop series on evolution in relation to neogene paleoclimates, since species are the units of variation which we use in discussing diversity in nature over time.

Much of evolutionary theory is undermined because it is based on an unsatisfactory species concept, the isolation concept (= biological species concept). In a long series of papers since 1973, I have pointed out the logical inconsistencies in the current genetical species concept.[1-6] Before discussing the implications of an alternative view of species for paleontology, I shall briefly summarize the problems facing the isolation concept.

GENETICAL SPECIES CONCEPTS

All current genetical concepts of species share a certain common ground in that they are compatible with the view[7] that a species can be seen in genetic terms as a field for gene recombination. However, disagreement exists over what determines the limits of this field and how the limits originate (see ref. 5 for a detailed discussion).

Isolation concept

The isolation concept is a strict version of the biological species concept freed from conflation with the recognition concept, which is discussed in detail below. Central to this concept is the fact that a species is

158

defined in terms of its reproductive isolation from other species. In other words, it is a relational concept, a fact seen by Mayr as an advantage. The isolation from other species is due to the action of isolating [p. 63] mechanisms, which are *ad hoc* characters with the function of preserving the genetic integrity of the species. Both Dobzhansky[11] (p. 104) and Mayr[12] (p. 284) explicitly regard speciation as an adaptive process "towards the most efficient utilization of the environment."

If we now return to Carson's view of species and ask what delineates the field for gene recombination according to the isolation concept, it is clearly the isolating mechanisms (ref. 10, p. 278). The shortcomings of the concept appear as soon as we attempt to answer the second question, "How do the isolating mechanisms arise initially?" Before attempting to do so, it should be noticed that the stated properties determine a number of logical consequences.

If the isolating mechanisms are indeed ad hoc characters, they are adaptations *s. str.*, as the appellation "mechanism" implies. This in turn means that they are the products of natural selection and, finally, that speciation must occur either in sympatry or in parapatry. Furthermore, if the function of the isolating mechanisms is really to protect the genetic integrity of a species, as Mayr has claimed, one should notice that the species is a population, a group, and, therefore, a character which has been selected to protect this group character must have been subject to group selection, not individual selection.

Drawing these consequences in this way makes us face up to what we are espousing. The fact that there is only one mode of speciation that seems to be universally accepted, geographic or allopatric speciation,[3,9,13-15] and that all other modes are hypothetical models, alone is enough to invalidate most of the expectations derived above.[16] Of course, Mayr's strong support for allopatric speciation is well known, and he even realizes the problem this poses for his view of species[9] (p. 548): "They are *ad hoc* mechanisms. It is therefore somewhat difficult to comprehend how isolating mechanisms can evolve in isolated populations." In fact his dilemma is evident from a careful study of ref. 9, pp. 548–53 (the equivalent problem confronts Dobzhansky,[8] p. 376). Neither author provided a satisfactory solution to the conflicts involved, simply because none is possible.

It is curious that Mayr should write (ref. 9, p. 551): "The thesis of the origin of reproductive isolation as a by-product of the total genetic reconstitution of the speciating population is consistent with all the known facts." If this were so, how can he believe that speciation is an adaptive process,[12] or that isolating mechanisms are ad hoc characters[9] which function to protect the genetic integrity of species, or that in the absence of close congeners isolating mechanisms can "afford" to be "general, non-specific, and variable" (p. 109)? If Mayr really accepted the view that isolating mechanisms arose through pleiotropy, they are clearly misnamed and should be renamed "isolating effects" *sensu* Williams.[17] Furthermore, much of Mayr's own theoretical writing will need revising, as it is inconsistent with the view that the isolating mechanisms are by-products of adaptation by members of an isolated population to a new environment. Thus, if a field of gene recombination is said to be delineated by isolating mechanisms, it is not at all enlightening to say that they arise pleiotropically, since this means, essentially, unpredictably.

Recognition concept
Faced with the logical difficulties of the isolation concept, a few of which have been given, the approach to species was rethought from the beginning,[1,5] including an examination[18] of the subliminal preconceptions which predispose us to thinking of species in relational terms.

If one accepts Carson's view of species, it is obvious that genetical concepts can apply only to organisms which are at least facultatively biparental, as in many aphid species. This means that sex is central to understanding species. Obligatory uniparental organisms are of course part of the diversity of life, but they follow a different genetic strategy from sexual organisms and cannot be forced under one heading for human minds more concerned with tidiness than comprehension of complex phenomena. Thus the key to genetical species is sex. Every biparental organism possesses a number of adaptations which comprise a system which leads to fertilization occurring. Fertilization is an event which will not be achieved fortuitously. Among the adaptations which lead to fertilization in motile biparental organisms there is a particularly critical subcategory comprising signal-response characters in the mating partners, the specific-mate recognition system (SMRS).[2,19] Well-studied examples of SMRS are those of the three- and ten-spined sticklebacks.[20] The SMRS plays a less critical role in achieving fertilization in primarily sessile organisms such as flowering plants or Mollusca such as oysters.[5] The raison d'être of the SMRS is to achieve effective fertilization under the conditions set by the organism's normal habitat, way of life (whether diurnal or nocturnal, motile or sessile, etc.) and during its normal breeding season. Although pleiotropy may play a role in shaping the SMRS, it ultimately is subject to selection: a signal influenced through pleiotropy will be subject to selection, just as it would be if influenced directly by mutation.

It is obvious that the characters of a fertilization system determine a field of gene recombination under natural conditions. Since the form of the fertilization system is clearly adaptive with respect to an organism's normal habitat, way of life, and normal breeding season, its characters are in the same category as other adaptive characters related to feeding, reproduction, and survival. Thus, it seems very clear that a new fertilization system, like any other adaptive character, might evolve under directional selection when an effectively small daughter population is isolated for sufficient time in a habitat different from the one to which the parental population is adapted. In many cases, of course, extinction is the outcome of such a restriction to a new habitat.

Once it is realized that the whole class of isolating mechanisms can be dispensed with entirely, the mind is freed from a set of ancient shackles[19]—true "tyrannies of the past"[21]—and genetic species can be looked at with fresh eyes. A genetic species is seen to be nothing more nor less than *that most inclusive population of biparental organisms which share a common fertilization system.*

Consequences should be drawn from this concept and used for attempts at falsification. Some of the more important are:

1. Species are not "adaptive devices," but are incidental consequences of adaptive evolution.

2. Speciation, therefore, cannot be regarded as "an adaptive process toward the most efficient utilization of the environment."[12] It is an incidental consequence of a small daughter population of sexual organisms adapting to a habitat different from the normal habitat of the parental population.

3. In contrast to the isolation concept, there are no problems with teleology involved in this concept.

4. This concept of species and speciation requires nothing more than individual selection.

5. The recognition concept is not a relational concept.

6. The fertilization system is very stable owing to three main factors: (i) The coadaptation of the male and female components of the SMRS; (ii) the fertilization system characters, including the SMRS characters, are closely related to the conditions of the normal habitat, and normal way of life, so, while these are acting, the fertilization system characters are under stabilizing selection; and (iii) large population size following the end of directional selection, etc.

7. The normal habitat is of fundamental importance, since it is involved in stabilizing species-specific characters such as the fertilization system. It follows that the fertilization system is

unlikely to change significantly while the organisms are in their normal habitat. Destabilization which encourages speciation will normally only follow the restriction of a small population to a different habitat type. Under these circumstances directional selection will replace stabilizing selection and lead to change, which [p. 64] will ultimately return to stabilizing selection.

8. Organisms are not restricted to particular places, but to particular habitats, their normal habitats; the organism's distribution changes with the expansion or contraction of the range of the habitat type, e.g., as ice sheets advance and recede with time.[22] It follows that speciation will be fostered not by such large-scale changes per se, but when small populations are trapped or restricted to new conditions.

9. Restriction to new conditions, for whatever reason, may lead to three possible outcomes: (i) extinction; (ii) subspeciation, as when the new conditions are not strongly divergent with respect to the normal habitat of the parental population; and (iii) speciation if the new conditions are such as to require significant adaptation to key fertilization system characters.

10. Since species are not "adaptive devices," hybridization takes on a new perspective. There is no idealistic specific integrity conceived; hybridization may have good or bad consequences which are subject to natural selection. Hybrids are often preserved as allopolyploids, for example.

SIGNIFICANCE OF SPECIES CONCEPTS FOR PALEONTOLOGY

Paleontologists have proved remarkably productive in their contributions to theoretical biology in the recent past. Their insights, added to those of the students of contemporary forms, are valuable indeed for the improved perspective they lead to. However, in the field of species biology it would seem to me that paleontology might make its greatest contribution as a testing ground for hypotheses generated by geneticists and other workers on the contemporary fauna and flora. One of the fundamental foundation stones of evolutionary and population biology is the species. This means that this foundation should be sound so that consequences of importance to paleontology can be drawn and acted on with confidence. An edifice built on an unsound foundation will also be unsound. I believe that the contemporary understanding of genetical species is unsound, which is why I have reexamined the fundamental genetical basis of species to provide a sounder alternative, largely free of the deficiencies facing the biological species concept. It remains to point out briefly some of the consequences for paleontology.

"Species" is a homonym, which is an immediate source of confusion. The same word is used in different ways. For example, it is used in two ways in taxonomy (the science of classification): as a basic category and for what are classified in that category (species taxa). Furthermore, species taxa may or may not coincide with another kind of species in population genetics, genetical species. Many contemporary species taxa represent, in actual fact, clusters of taxonomically undetected genetical species. Paleontologists are virtually unable to accommodate cryptic species in their work for obvious reasons. However, there is every reason for them to be aware that they exist and to take this inevitable difficulty into account in their theoretical work. To treat fossil taxonomic species as genetical species needs to be done with due attention to the problems involved (i.e., the distortion thus introduced). (Cryptic genetical species, i.e., sibling species, are typical genetical species which use nonvisual signals in their SMRS, often because they are nocturnal.)

In this article and in others cited, I have shown clearly the difficulties and contradictions that arise from adopting a relational genetical concept of species as is done at present. A far more acceptable approach is that provided under the recognition concept of species. Besides being inconsistent and providing impossible philosophical conflicts, the current isolation concept generates predictions for paleontology different from the recognition concept (Table 1).

If one were to ask, What pattern would one expect to observe in fossil lineages if the recognition concept were correct? one would receive a clear answer: A pattern similar to that which has become known as the punctuated equilibrium pattern.[23] While both the recognition and isolation concepts might predict the "punctuation" aspect, there is nothing inherent in the isolation concept which would predict the "equilibrium" aspect. Maynard Smith[24] has commented, "As a geneticist I was left with the impression that the 'sudden' appearance of a new species called for no new explanations, but that stasis may do." It has long been emphasized[2–6] that the coadaptation of SMRS "signals" and "receivers" of mating partners, the stabilizing selection imposed on organisms by their restriction to a preferred, species-specific habitat type, and the very size of most natural populations, imposes a very real stability on species—a sort of homeostasis. This has previously been noticed by Mayr[10] (p. 300), but on rather vague grounds. Henderson and Lambert[25] have demonstrated the uniformity of the fertilization system in the cosmopolitan species *Drosophila melanogaster*.

Table 1. Some consequences derived from the recognition concept of species which are relevant to paleontology

---

1. Speciation is allopatric.
2. Speciation involves effectively small populations.
3. Species are incidental consequences of adaptive evolution.
4. Species-specific characters are acquired at speciation.
5. Species are stable due to homeostatic effects: coadaptation of signal and response characters in the mating partners; adaptation of adaptive characters, including fertilization characters, to environment of preferred habitat; large population size achieved after speciation.
6. Speciation is one of four possible outcomes of isolating a small population in a new environment: no essential change, extinction, subspeciation, speciation.
7. Habitat preference is due to the fixation at speciation of species-specific characters.
8. Species are not defined relationally.
9. Members of specialist species are more likely to speciate or become extinct than are members of generalist species.

The fundamentally important role of the normal or preferred habitat (assumed to approximate to the habitat type in which the species first originated) is central to the recognition concept, but not the isolation concept. The way habitat type determines the distribution of species has been dramatically shown by the studies of Coope,[22] work which should be considered carefully by anyone taking the Red Queen hypothesis of Van Valen[26] seriously.[4]

It is significant to notice that the recognition concept predicts that speciation follows the isolation of a small population in a habitat type different from that preferred by the parental population, and that it is this change that destabilizes the homeostatic properties of the species-specific characters of the original population. While in their preferred habitat type, members of a species are highly stabilized. This means that even isolated daughter populations are unlikely to speciate *provided the daughter population is restricted to an environment like that of the parental population,* that is, to a habitat approximately the same as the one in which the species evolved. Restriction to a new habitat type can occur in a number of ways. A small number of birds might be blown from a mainland to a rather distant island where the parental habitat type is not represented; or, on a continent, interpluvial conditions may lead to the long-term isolation of forest species in, first, relic patches of forest but, ultimately, force them into a savanna habitat through the disappearance of the trees. This scenario is probably widely valid in Africa and may indeed be the basis of the displacement of man's own lineage from forest to grassland. The ground woodpecker (*Geocolaptes olivaceus*) is another species that would fit this model.

The recognition concept does not assign a major driving role to interspecific competition.[27] Displacement to new habitats occurs during speciation and is not due to competitive interactions, as has so often and so glibly been postulated in the past. Similarly, character displacement is considered unlikely.[28]

Enough has been written to demonstrate the utility of arguing from a detailed genetical species concept. Perhaps one final point should be emphasized, that there is a very basic difference between the geographic speciation model and the one briefly outlined here. In fact, all they have in common is that speciation occurs in allopatry. In the case of Mayr's model, speciation is supposed to result from the so-called isolating mechanisms being modified through pleiotropic change during adaptive evolution; in the other case, the fertilization system is adapted to a new habitat type so that signaling between the mating partners is restored to full effectiveness under the new conditions. Once directional selection has adjusted the characters of the fertilization system, the characters will once more be under stabilizing selection. It is only at this point that the crisis of speciation is over and the population is able to grow and expand. This is why speciation is a cryptic event as far as the paleontological record is concerned. Under Mayr's view of speciation, there is no relationship between the form of isolating mechanisms and the habitat of the species, whereas under the alternative recognition concept there is [p. 65] obviously a very close relationship between the fertilization system and the habitat type, as well as a relationship to the organism's way of life and breeding season. The fact that such relationships exist is very obvious.[5] Thus, the points listed in Table 1 set new predictions of use to paleontologists in interpreting the fossil record. Furthermore, they provide a basis for testing the recognition concept and possibly falsifying it.

REFERENCES

1. Paterson, H.E.H. 1973. Animal species studies. Pp. 31–36 in Animal and plant speciation studies in Western Australia, by H.E.H. Paterson and S. H. James. *Journal of the Royal Society of Western Australia* 56:31–43. [See this volume, Author's Preface.]
2. Paterson, H.E.H. 1978. More evidence against speciation by reinforcement. *South African Journal of Science* 74:369–71. [This volume, chap. 3.]
3. Paterson, H.E.H. 1981. The continuing search for the unknown and unknowable: A critique of contemporary ideas on speciation. *South African Journal of Science* 77:113–19. [This volume, chap. 5.]

4. Paterson, H.E.H. 1982. Perspective on speciation by reinforcement. *South African Journal of Science* 78:53–57. [This volume, chap. 8.]

5. Paterson, H.E.H. 1985. The Recognition Concept of Species. In *Species and Speciation*, ed. E. S. Vrba, 21–29. Transvaal Museum Monograph No. 4. Pretoria: Transvaal Museum. [This volume, chap. 12.]

6. Lambert, D. M., and H.E.H. Paterson. 1984. On "Bridging the gap between race and species": The isolation concept and an alternative. *Proceedings of the Linnean Society of New South Wales* 107:501–14.

7. Carson, H. L. 1957. The Species as a Field for Gene Recombination. In *The Species Problem*, ed. E. Mayr, 23–38. Publication No. 50. Washington, D.C.: American Association for the Advancement of Science.

8. Dobzhansky, T. 1970. *Genetics of the Evolutionary Process.* New York: Columbia University Press.

9. Mayr, E. 1963. *Animal Species and Evolution.* Cambridge, Mass.: Belknap Press of Harvard University Press.

10. Mayr, E. 1970. *Populations, Species and Evolution.* Cambridge, Mass.: Belknap Press of Harvard University Press.

11. Dobzhansky, T. 1976. Organismic and Molecular Aspects of Species Formation. In *Molecular Evolution*, ed. F. J. Ayala, 95–105. Sunderland, Mass.: Sinauer Associates.

12. Mayr, E. 1949. Speciation and Systematics. In *Genetics, Paleontology, and Evolution*, ed. G. L. Jepsen, E. Mayr, and G. G. Simpson, 281–98. Princeton, N.J.: Princeton University Press.

13. Dobzhansky, T. 1951. *Genetics and the Origin of Species.* 3d ed. New York: Columbia University Press.

14. Bush, G. L. 1975. Modes of animal speciation. *Annual Review of Ecology and Systematics* 6:339–64.

15. White, M.J.D. 1978. *Modes of Speciation.* San Francisco: W. H. Freeman.

16. Paterson, H.E.H., and M. Macnamara. 1984. The recognition concept of species (Michael Macnamara interviews H.E.H. Paterson). *South African Journal of Science* 80:312–18. [This volume, chap. 10.]

17. Williams, G. C. 1966. *Adaptation and Natural Selection.* Princeton, N.J.: Princeton University Press.

18. Paterson, H.E.H. 1982. Darwin and the origin of species. *South African Journal of Science* 78:272–75. [This volume, chap. 9.]

19. Paterson, H.E.H. 1980. A comment on "Mate Recognition Systems." *Evolution* 34:330–31. [This volume, chap. 4.]

20. Morris, D. 1970. *Patterns of Reproductive Behaviour.* London: Jonathan Cape.

21. Robinson, J. H. 1926. *The Mind in the Making.* London: Jonathan Cape.

22. Coope, G. R. 1975. Late Cenozoic fossil Coleoptera: Evolution, biogeography, and ecology. *Annual Review of Ecology and Systematics* 10:247–67.

23. Eldredge, N., and S. J. Gould. 1972. Punctuated Equilibria: An Alternative to Phyletic Gradualism. Chap. 5 in *Models in Paleobiology*, ed. T. J. M. Schopf, 82–115. San Francisco: Freeman, Cooper and Company.

24. Maynard Smith, J. 1981. Macroevolution. *Nature* 289:13–14.

25. Henderson, N. R., and D. M. Lambert. 1982. No significant deviation

from random mating of worldwide populations of *Drosophila melanogaster. Nature* 300:437–40.

26. Van Valen, L. 1973. A new evolutionary law. *Evolutionary Theory* 1:1–30.

27. Walter, G. H., P. E. Hulley, and A.J.F.K. Craig. 1984. Speciation, adaptation and interspecific competition. *Oikos* 43:246–48.

28. Grine, F. E. 1981. Trophic differences between "Gracile" and "Robust" Australopithecines: A scanning electron microscope analysis of occlusal events. *South African Journal of Science* 77:203–30.

14 On Defining Species in Terms of Sterility: Problems and Alternatives

Paterson, H.E.H. 1988. On defining species in terms of sterility: Problems and alternatives. *Pacific Science* 42:65–71.

A problem with attempting to overthrow a well-established concept, such as the isolation concept, is that it is almost inevitable that one should repeat oneself many times. Every audience one addresses is likely to be unfamiliar with the new ideas and will require at least an outline of what one is saying, and how it compares with the existing view. Thus, even with the distinguished audience which gathered at Honolulu in early June 1985 to honor Hampton L. Carson, one could not assume familiarity with the recognition concept. The theme I chose was to consider the role of sterility (considered in broad terms) in defining species. This theme has been touched on in my 1982 paper (chap. 9) and in the 1985 overview (chap. 12).

The chairman of my session was none other than Ledyard Stebbins, who informed the audience that he had waited a long time to come face-to-face with me, an unmitigated zoological chauvinist, who paid no attention to plants (i.e., angiosperms!). As it happened, I had noticed who my chairman would be and had used a number of plant examples and dealt with topics like polyploidy, which are dear to plant cytogeneticists' hearts.

In this paper I took the opportunity to emphasize that the recognition and isolation concepts diverge sharply over the significance of sterility in defining species. It is on this matter that many part company with me (see, for example, Butlin's 1987 views). In particular, many find it hard to accept my view that autopolyploidy is an intraspecific change.

Analogously, Clarkia lingulata *seems to have arisen from C.* biloba *by extensive chromosome repatterning. Both share a common fertilization system, attracting the same pollinators. When crossed they are, predictably, intersterile.*

> *Under these circumstances the two "species" cannot coexist;*
> *the rarer form will soon be eliminated, as pointed out in my*
> *1978 paper (chap. 3). Lewis has shown this to be the case in*
> *his areas of experimental sympatry. This is indeed unexpected*
> *with two good species.*
>
> *Speaking of "reproductive isolation" seems contradictory*
> *when matings occur at random between two populations in*
> *sympatry, even if they are intersterile.*
>
> *Ernst Mayr was a referee of this paper, which he did not*
> *like, but still recommended for publication.*

ABSTRACT

Despite its historic role as a criterion of species status, intersterility *sensu lato* is not an acceptable characteristic for delineating the genetic species or field of gene recombination. This conclusion is not new since it is in agreement with Darwin's views as expressed in the *Origin of Species* (1859). The critical role of sterility in distinguishing between the prevailing genetic concept of species and its rival, the recognition concept, is demonstrated. Factors that may have led to the general acceptance of Wallace's views on speciation, rather than Darwin's, are briefly discussed.

Perhaps because of ancient knowledge of the mule, people in general have long considered sterility *sensu lato* as the key to the delineation of species. In Western societies this belief was held first by articulate Christians and then by Christian biologists, in support of their pre-conceptions that were founded on the creation stories of the book of Genesis. Eventually, this belief spread by cultural osmosis to biologists in general, although they often had no commitment to the Biblical accounts of creation. Darwin attested to this situation in the opening sentence of chapter 8 of the *Origin of Species* (1859:245): "The view generally entertained by naturalists is that species, when intercrossed, have been specially endowed with the quality of sterility, in order to prevent the confusion of all organic forms." The historian Ellegard (1958:208) supported Darwin's claims with evidence from the popular and ecclesiastical journals of the 1850s. Well before this, the great geologist and committed Christian, Lyell, wrote (1832:19), "Nature has forbidden the intermixture of the descendants of distinct original stocks, or has, at least, entailed sterility on their offspring, thereby preventing their being confounded together, and pointing out that a multitude of distinct types must have been created in the beginning, and must have remained pure and uncorrupted to this day." Mayr's (1963:109) views of almost a century and a half later bear a strong generic resemblance to Lyell's: "It is the function of the isolating mechanisms [which include sterility] to prevent such a breakdown [due to hybridization] and to protect the integrity of the genetic system of species." The two authors share the view that sterility is an isolating mechanism. Lyell saw the function of sterility as the protection of a species' divine, and Mayr its genetic, integrity.

While A. R. Wallace and T. H. Huxley both subscribed to this view, Darwin's ideas were strikingly heterodox. The opening to chapter 8 of the *Origin of Species* (1859:245) continues:

> This view certainly seems at first probable, for species within the same country could hardly have kept distinct had they been capable of crossing freely. The importance of the fact that hybrids are very generally sterile, has, I think, been much underrated by some late writers. On the theory of natural selection the case is especially important, inasmuch as the sterility of hybrids could not possibly be of any advantage to them, and therefore could not have been acquired by the continued preservation of successive profitable degrees of sterility. I hope however, to be able to show that sterility is not a specially acquired or endowed quality, but is incidental on other acquired differences.

There is no hint here of a function for sterility as an isolating mechanism.

When we define adaptation strictly [p. 66] (Williams, 1966), Lyell's and Mayr's statements clearly show that these authors have regarded sterility as an adaptation, while Darwin considered it to be an incidental effect of adaptive change.

One of the problems facing anyone attempting to understand Mayr in depth is that he is inconsistent, as was Dobzhansky, and as are many other evolutionists. It is quite possible to find that he has also espoused Darwin's viewpoint that sterility is an "effect" not an "adaptation" (mechanism) (e.g., Mayr, 1963:551) and that sterility cannot be used to delineate species (Mayr, 1942:119). Elsewhere, however, Mayr advocates "instantaneous speciation," which entails a definition of species in terms of sterility and a belief that a new species can arise as one individual (Mayr, 1970:254). I have therefore been obliged to emphasize the view most consistent with his species definitions and his statements on the ad hoc nature of isolating mechanisms; these are worded quite unambiguously in many places, for example, in Mayr (1963:20,91,109,129,548). I appreciate that I am attempting a difficult task because I am certain to be accused of distorting the author's views, but this is a hazard in attempting to understand an author's viewpoint when his writings are extensive and inconsistent. It should be made clear that it is the detection of inconsistencies with the current paradigm that has driven me to put forward an alternative view of species that is less inconsistent and yet in keeping with the critical data from observation and experiment (see Paterson, 1985 [chap. 12], for more details).

The object of this paper is to reexamine sterility as a basis for defining species. This can be done rationally only within the constraints of a particular genetic concept of species, for sterility is seen differently under different genetic concepts of species.

SPECIES IN GENETIC TERMS

Carson (1957) pointed out that species are fields for gene recombination, and I think that few would contest this point. Disagreement exists over how the field for gene recombination is delineated. Two alternative suggestions have been made. One is the isolation concept: The field is determined by a diverse set of characters, the isolating mechanisms, the function of which is to preserve the integrity of the genetic system of the species (Mayr, 1963:109, 1982:272). The other is the recognition concept: The characters of a species that function to bring about fertilization in a population's normal habitat automatically delimit the field for gene recombination. Earlier writers have consistently failed to notice that these ways of looking at species are conceptually quite distinct. This results in conflation occurring more commonly than even Verne Grant noticed (Grant, 1971:35). Thus, the isolation concept is the heart of the biological species concept of Dobzhansky and Mayr, freed from conflation with the recognition concept.

The isolation concept is a relational concept, one species being defined in relation to another. This was seen by Mayr (1963:20) as an advantage: "Species are more unequivocally defined by their relation to nonconspecific populations ('isolation') than by the relation of conspecific individuals to each other. The decisive criterion is not the fertility of individuals but the reproductive isolation of populations." This emphasis (which distinguishes the isolation concept from the recognition concept) is crucial to the consideration of sterility as a basis for delineating species.

STERILITY AND THE ISOLATION CONCEPT

It has long been known that sterility *s. lat.* is an unsatisfactory criterion for delineating the limits of a field for gene recombination (Darwin, 1859:245ff; Mayr, 1942:119). I shall therefore restrict discussion to aspects of the matter that are critical for the isolation concept.

Mayr (1970:12) wrote, "Species are groups of interbreeding natural populations that are reproductively isolated from other such groups." It is fundamental to the assessment of the isolation concept of species to understand the nature and origin of the [p. 67] isolating mechanisms on which the reproductive isolation (i.e., delineation of the field for gene recombination) depends. The term "mechanisms" implies that they are adaptations (Crowson, 1970:221; Futuyma, 1979:408; Williams, 1966:9). This in turn implies that they have been "fashioned by selection for the goal" of "protecting" the "genetic integrity" of a species. It is my view that this inappropriate use of the term "mechanism" has been a serious disadvantage to evolutionary theory through the loss of clarity of thought.

The generally recognized isolating mechanisms have been classified by Mecham (1961) as either premating or postmating mechanisms (Mayr, 1963:92). This at once demonstrates their fundamental heterogeneity, because postmating mechanisms cannot originate under selection, while premating mechanisms conceivably can. For this reason alone it is inconsistent to call sterility *s. lat.* an isolating mechanism. Since sterility is an unsatisfactory basis for delineating species (Mayr, 1942:119), and since postmating "mechanisms" are fundamentally different from premating mechanisms, they are obviously out of place in a table of isolating mechanisms. How can we discuss intelligently the nature and origin of isolating mechanisms if we include totally extraneous elements among them (Mayr, 1963:91)?

These objections to sterility as sufficient grounds for delimiting a field for gene recombination have important consequences. When a tetraploid angiosperm arises within the range of a diploid species, this is unequivocally "instantaneous speciation" according to Mayr (1970:254). This opinion has long been a source of disagreement between geneticists and plant taxonomists who, not surprisingly, are reluctant to name as distinct species-taxa organisms that may be indistinguishable structurally from their diploid relatives (Cronquist, 1978). Mayr and others who accept the isolation concept regard the tetraploid (individual, not population!) as being reproductively isolated from the parental diploid individuals, because the offspring of any cross between the two will yield triploid progeny that are more or less sterile due to meiotic difficulties. (In fact, in population genetics terms, s is seldom unity [Lewis, 1967].)

This is actually not a satisfactory scenario as it stands; it requires a good deal of propping up before it begins to look like a speciation model. If the polyploid and diploid plants are committed to outbreeding, the newly arisen tetraploid will simply be eliminated (Li, 1955; Paterson, 1981 [chap. 5]). Subsidiary propping up is needed. Usually, this is in the form of the tetraploid persisting through alternative reproductive strategies (e.g., by vegetative reproduction or self-fertilization). But, of course, uniparental organisms are not covered by any of the genetic concepts of species. In a biparental species, the newly arisen tetraploid will be eliminated, as has already been emphasized. Furthermore, Lewis (1967) has pointed out that gene exchange is not completely interrupted between tetraploid and diploid individuals. Finally, what does gene exchange in such a case mean? The "gene pool" of the tetraploid is a subset of the gene pool of the diploid population. The only consequence that could result from total reproductive isolation of the two populations is that they could eventually evolve independently. This, however, would be more or less prevented by even quite a low gene flow between them.

It might be argued: But in nature one observes genera within which the species differ in their levels of ploidy; clearly polyploidy has been the major cause of speciation here. How clearly is this established? Such arrays of congeneric species do not establish that polyploidy led directly to their origination as species. With this in mind, consider the following points: How do we know that the congeneric species of the array did not arise under the conditions of allopatric speciation as other species do? Perhaps the only role that polyploidy has (regardless of whether we are considering allopolyploidy or autopolyploidy) is that it makes sympatry between diploids and polyploids impossible. Polyploid seeds dispersed into different allopatric habitats, to which they may well be more suited, could then speciate by adapting to the new conditions in the usual way. The [p. 68] usual argument involves the old fallacy of inferring causality from a correlation. It is the same fallacy that led White (1978) and others to infer causality for speciation from observing that two congeneric species differ by a fixed gene rearrangement. As Carson et al. (1967) and others have observed, such correlations with speciation are adequately explained by understanding that speciation involves small populations, which, in turn, favor the fixation of such rearrangements.

These few points are sufficient to call seriously into question the conventional views on speciation via polyploidy, as well as Mayr's views on sterility as a "fully efficient" isolating mechanism that would enable two populations (diverged in allopatry) to return to sympatry where "reinforcement" could occur (Mayr, 1963:551).

Sterility occurs among plants for other reasons than from polyploidy. A much cited example should perhaps receive further consideration: *Clarkia biloba* / *C. lingulata* (Lewis, 1953; Lewis and Raven, 1958). In this case, *C. lingulata* evidently arose through extensive chromosome repatterning in a small, isolated population (Lewis, 1966). Translocation experiments (Lewis, 1961) provided support for the view outlined above that sterility is not an efficient primary isolating mechanism and supported the prediction from population genetics theory that the rarer population will generally soon be eliminated. Lewis (1961) provided evidence that pollinators do not distinguish between *C. lingulata* and its parental form. At present, there appear to be two small, isolated populations of *C. lingulata* surviving in allopatry in much the way that was suggested for polyploids while they speciate allopatrically. Thus, in this view, neither the occurrence of polyploidy nor chromosomal repatterning as bases for crossing sterility constitutes rapid speciation, but is merely a factor possibly leading to speciation in allopatry. This topic will be considered further in the next section.

Darwin (1859) drew attention to the fact that interfertility often exists between undoubted species. Since then, many other examples have become known (Mayr, 1963:90). A widely quoted example (Mayr, 1963:90) is that of the interfertility of the common sympatric duck species, the mallard (*Anas platyrhynchos*) and the pintail (*A. acuta*), two species which are not even very closely related. Darwin was fully aware of this case and provided many other examples (Stauffer, 1975). Darwin (1859:250) also drew attention to the fact that sometimes intraspecific crosses are less fertile than some interspecific crosses involving the same species. It is conventional to talk of the sterility of crosses involving flowers on the same plant as *incompatibility*, but one should be conscious of the fact that to do so may be to mislead, since, in some cases at least, no new phenomenon is involved (Heslop-Harrison et al., 1974). Accordingly, the practice should perhaps be reviewed since it could allow us to persist with our preconceptions. It is not easy to see how a self-incompatibility system could be selected to function as an outbreeding device, for self-incompatibility is obviously generally more disadvantageous than the possible disadvantages of inbreeding depression. On the other hand, selection for outbreeding devices such as distyly is conceivable as a device to counter preexisting, fortuitously disadvantageous "self-incompatibility."

As has already been made clear, it is also impossible to evolve a system of interspecific incompatibility through natural selection. One wonders why more attention has not been paid to these points by workers on pollination mechanisms, since they would provide a new perspective to their work.

Sterility has a special importance under the isolation concept, because it can obviously be acquired in allopatry, and is thus compatible with allopatric speciation. It can provide the basis for selection for true "isolating mechanisms" should diverged populations return to sympatry (Mayr, 1963:551). Although Mayr has always strongly supported allopatric speciation, he was aware that it posed problems to anyone who believed in speciation as an adaptive process: "They [isolating mechanisms] are ad hoc mechanisms. It is therefore somewhat difficult to comprehend how isolating mechanisms can evolve in isolated populations" (Mayr, 1963:548). However, it is curious that [p. 69] he should have ignored the fact that reinforcement is really not to be expected when two diverged populations with hybrids of reduced fitness meet; what would be expected under these conditions is that natural selection would act to eliminate the cause of sterility, or the less common population if sterility is absolute (i.e., $s = 1$) (Lambert et al., 1984 [chap. 11]; Paterson, 1978 [chap. 3]). If an author were to adopt an extreme position and advocate that all "isolating mechanisms" evolve in isolation by pleiotropy, they could not then be called "mechanisms" as they are obviously effects. It is then not possible to believe that speciation is an adaptive process, as did Mayr (1949:284) and Dobzhansky (1976:104), and "isolating mechanisms" can no longer be treated as ad hoc characters. Perhaps enough has been said to support and extend the early view of Mayr (1942) that sterility is not in itself satisfactory as the basis for defining species, even when reproductive isolation is the criterion for species, regardless of whether authors list it or its variants as "isolating mechanisms."

STERILITY AND THE RECOGNITION CONCEPT

Sterility under the recognition concept of species takes on a different aspect entirely. Paterson (1982a [chap. 8], 1985 [chap. 12]) has defined as a species the most inclusive population of individual biparental organisms that share a common fertilization system. The fertilization system of a species comprises all characters that contribute to the achievement of fertilization. These characters are diverse and include such characters in the mating partners as the design features of the gametes, those determining synchrony in the achievement of reproductive condition, the coadapted signals and receivers of mating partners, and their coadapted organs of gamete delivery and reception.

The adaptations of the fertilization system act efficiently under the conditions of the normal habitat and way of life of the organisms. (Way-of-life characters are those such as nocturnal versus diurnal activity cycles, sessility versus motility, etc.) The fertilization system leads to positive assortative mating among members of a population in its normal habitat. This is the same as saying that the fertilization system determines the limits of the field for gene recombination in nature, and it does so without any reference to any other field for gene recombination. The recognition concept is not a relational concept (Paterson, 1985 [chap. 12]). Thus, speciation in geographic isolation poses for it none of the conceptual problems that it does for the isolation concept. The consequence of all this is that species are incidental results of adaptational and stochastic change in isolated, small populations. Adaptation of the fertilization system to the new conditions occurs just the same way as do all other adaptive characters; no special reliance on pleiotropy is involved.

Applying this concept to the study of sterility *s. lat.* leads to quite different insights from the isolation concept. Let us examine the case of *Clarkia biloba* and *C. lingulata* in this light. Lewis (1961) states explicitly that pollinators do not distinguish between these two populations when they are brought into sympatry by geneticists and that crosses between them are intersterile, although the *C. lingulata* gene pool is probably a mere subset of that of *C. biloba*. The sterility is due to problems with meiosis caused by the complex chromosomal rearrangements that distinguish the two populations. As far as these data go, they indicate quite clearly that the two populations share a common fertilization system and are conspecific. According to the predictions of population genetics algebra (Li, 1955), such populations should not be able to coexist, and the least abundant population should be eliminated rather quickly. Transplantation studies by Lewis (1961) support this expectation, with three out of the four populations conforming within the five generations spanned by the study. Harper and Lambert (1983) have carried out a uniquely planned and careful experimental study of such situations, which are also in agreement with Lewis's observations. Computer simulations point the same way (Lambert et al., 1984 [chap. 11]).

To summarize this part: Obviously, sterility cannot be among the adaptations to [p. 70] bring about effective fertilization (the fertilization mechanisms); it is clearly not a relevant factor in delineating the field for gene recombination of a species. Thus, the diverse forms of intersterility (including those due to autopolyploidy, self-incompatibility in monoecious plants, and "cytoplasmic incompatibility" [Laven, 1967] as found in *Drosophila paulistorum* and the *Culex pipiens* complex) are regarded as intraspecific phenomena. This conclusion is in sharp contrast to interpretations based on the isolation concept and is an answer to those who believe the two concepts to be "opposite sides of the same coin."

DISCUSSION

Sterility *s. lat.* is of central significance for evolutionary theory because of the ancient association of sterility with interspecific crossing exemplified by the mule. Sterility was thus first conceived as an "isolating mechanism," filling a role required by two separate and independent commitments of Western society, the first, a commitment to purity of lineage required by the practices of ancient animal and plant breeders and the second, a commitment to the creation stories of Christianity and Judaism, which required the preservation of the Creator's handiwork. With this existing, subliminal, cultural bias, it is scarcely surprising that Wallace's views on the nature of species were preferred to Darwin's (Paterson, 1982b [chap. 9]). We are told (Mayr, 1963) that Darwin eliminated the species as a concrete natural unit, and thereby neatly eliminated the need for a solution to the problem of how species multiply. In fact, Darwin conceived species much as many taxonomists and others, in practice, do today. Furthermore, he did discuss the origin of species under natural selection (Darwin, 1859:104–5). What Darwin saw—and Wallace did not see—was that species are not adaptive devices, but incidental products of adaptive evolution. In this he differed not only from nearly all his philosophical predecessors, but also from A. R. Wallace, J. T. Gulick, G. C. Robson, R. A. Fisher, T. Dobzhansky, and Ernst Mayr (Paterson, 1982b [chap. 9]). This unique contribution of Darwin's is probably his most revolutionary and is the one that has scarcely been noticed except by philosophers (Hull, 1973:56; Kuhn, 1970:172).

Sterility focuses our attention sharply on the importance of adhering to a concept of species that is logically consistent. It is not a matter of fashion or convenience, but a vital matter of basic comprehension. As Darwin said 126 years ago, "The case is especially important," not only for the theory of natural selection, but for understanding the genetic nature of species.

ACKNOWLEDGMENTS

This contribution is dedicated to Hampton L. Carson, in admiration of his leadership in the field of evolutionary genetics and appreciation of his friendship and encouragement.

I thank Judith Masters for drawing my attention to Lyell's views on sterility.

178   *Evolution and the Recognition Concept of Species: Collected Writings*

REFERENCES

Carson, H. L. 1957. The Species as a Field for Gene Recombination. In *The Species Problem*, ed. E. Mayr, 23–38. Publication No. 50. Washington, D.C.: American Association for the Advancement of Science.

Carson, H. L., F. E. Clayton, and H. D. Stalker. 1967. Karyotypic stability and speciation in Hawaiian *Drosophila*. *Proceedings of the National Academy of Science, USA* 57:1280–85.

Cronquist, A. 1978. Once Again, What Is a Species? In *Biosystematics in Agriculture*, ed. J. A. Romberger, 3–20. New York: Wiley.

Crowson, R. A. 1970. *Classification and Biology*. London: Heinemann.

Darwin, C. 1859. *On the Origin of Species by Means of Natural Selection; or, the Preservation of Favoured Races in the Struggle for Life*. London: John Murray.

Dobzhansky, T. 1976. Organismic and Molecular Aspects of Species Formation. In *Molecular Evolution*, ed. F. J. Ayala, 95–105. Sunderland, Mass.: Sinauer Associates.

Ellegard, A. 1958. Darwin and the General Reader. Gothenburg Studies in English. Vol. 8. Göteborg, Sweden: University of Göteborg.

Futuyma, D. J. 1979. *Evolutionary Biology*. Sunderland, Mass.: Sinauer Associates.

Grant, V. 1971. *Plant Speciation*. New York: Columbia University Press.

Harper, A. A., and D. M. Lambert. 1983. The population genetics of reinforcing selection. *Genetica* 62:15–23.

Heslop-Harrison, J., R. B. Knox, and Y. Heslop-Harrison. 1974. Pollen-wall proteins: Exine-held fractions associated with the incompatibility response. *Theoretical Genetics* 44:133–37.

Hull, D. 1973. *Darwin and His Critics*. Cambridge, Mass.: Harvard University Press.

Kuhn, T. 1970. *The Structure of Scientific Revolutions*. 2d ed. Chicago: University of Chicago Press.

Lambert, D. M., M. R. Centner, and H.E.H. Paterson. 1984. Simulation of the conditions necessary for the evolution of species by reinforcement. *South African Journal of Science* 80:308–11. [This volume, chap. 11.]

Laven, H. 1967. Speciation and Evolution in *Culex pipiens*. In *Genetics of Insect Vectors of Disease*, ed. J. W. Wright and R. Pal, 251–75. Amsterdam: Elsevier.

Lewis, H. 1953. The mechanism of evolution in the genus *Clarkia*. *Evolution* 7:1–20.

———. 1961. Experimental sympatric populations of *Clarkia*. *American Naturalist* 95:155–68.

———. 1966. Speciation in flowering plants. *Science* 152:167–72.

———. 1967. The taxonomic significance of autopolyploidy. *Taxon* 16:267–71.

Lewis, H., and P. H. Raven. 1958. Rapid evolution in *Clarkia*. *Evolution* 12:319–36.

Li, C. C. 1955. *Population Genetics*. Chicago: University of Chicago Press.

Lyell, C. 1832. *Principles of Geology*. Vol. 2. London: Murray.

Mayr, E. 1942. *Systematics and the Origin of Species.* New York: Columbia University Press.

———. 1949. Speciation and Systematics. In *Genetics, Paleontology, and Evolution,* ed. G. L. Jepsen, E. Mayr, and G. G. Simpson, 281–98. Princeton, N.J.: Princeton University Press.

———. 1963. *Animal Species and Evolution.* Cambridge, Mass.: Belknap Press of Harvard University Press.

———. 1970. *Populations, Species and Evolution.* Cambridge, Mass.: Belknap Press of Harvard University Press.

———. 1982. *The Growth of Biological Thought.* Cambridge, Mass.: Harvard University Press.

Mecham, J. S. 1961. Isolating Mechanisms in Anuran Amphibians. In *Vertebrate Speciation,* ed. W. F. Blair, 24–61. Austin: University of Texas Press.

Paterson, H.E.H. 1978. More evidence against speciation by reinforcement. *South African Journal of Science* 74:369–71. [This volume, chap. 3.]

———. 1981. The continuing search for the unknown and unknowable: A critique of contemporary ideas on speciation. *South African Journal of Science* 77:113–19. [This volume, chap. 5.]

———. 1982a. Perspective on speciation by reinforcement. *South African Journal of Science* 78:53–57. [This volume, chap. 8.]

———. 1982b. Darwin and the origin of species. *South African Journal of Science* 78:272–75. [This volume, chap. 9.]

———. 1985. The Recognition Concept of Species. In *Species and Speciation,* ed. E. S. Vrba, 21–29. Transvaal Museum Monograph No. 4. Pretoria: Transvaal Museum. [This volume, chap. 12.]

Stauffer, R. C. 1975. *Charles Darwin's Natural Selection.* Cambridge: Cambridge University Press.

White, M.J.D. 1978. *Modes of Speciation.* San Francisco: W. H. Freeman.

Williams, G. C. 1966. *Adaptation and Natural Selection.* Princeton, N.J.: Princeton University Press.

15 A View of Species

Paterson, H.E.H. 1989. A View of Species. Chap. 6 In
Dynamic Structures in Biology, ed. B. Goodwin, A. Sibatani,
and G. Webster, 77–88. Edinburgh: Edinburgh University
Press.

*"A View of Species" was presented at Osaka at the First
International Workshop on Structuralism in Biology in
December 1986. This work is essentially an essay explaining
to biologists who are not evolutionists the significance of
species and speciation for understanding the evolution of
natural species diversity. In it I emphasized the significance
of symbiosis in evolutionary theory, pointing out its neglect in
major evolutionary works. I said a little more about sexual
selection in this essay than previously. The importance of the
"way-of-life" character, sessility, was again emphasized (I
had made the distinction in my 1985 paper [chap. 12]).*

*I think the importance of this paper is due to its essay
form, which makes the ideas of the recognition concept more
accessible to general biologists. Because it was directed at a
new audience, evolutionists will find this chapter rather
familiar.*

My predilection in biology is to search for an effective evolutionary
synthesis and a credible biological world view. Though not consciously
a structuralist, I appear to share aspirations and goals with at least
some structuralists. For example, I feel empathy with the structuralists' concern for wholeness, the ideas of transformation and self-
regulation, if I have conceived these abstract ideas in Piaget's way.

Evolution, particularly species theory, is a significant subject because it illuminates much of what we see in the living world and our
own species. Such illumination can have far-reaching consequences
not only in population biology, but for the way we see ourselves, as
became obvious when society first realized that the human state was
not the ultimate creation of a deity but the product of a process which
had yielded millions of other such products, each remarkable in some

way. This was a revolution in human thought appreciated by rather few other than philosophers. The pursuit of such an intellectual path is only possible for members of *Homo sapiens*, a fact which helps identify a more particular evolutionary perplexity, the achievement by humans of the conceptual powers which enable such a problem to be recognized.

In this chapter I shall consider from perspectives which may interest structuralists a number of evolutionary topics related to natural diversity, species, and the origin of species. Although I retain common ground with the movement which led to the so-called modern synthesis of evolutionary thought, I do not subscribe to a number of ideas and viewpoints which currently receive general support. My aim is to present a heterodox ordering of well-known information in order to provide a new perspective, rather than new facts. In particular, I wish to emphasize the fundamental importance of genetical species in evolution, despite their origins as serendipitous consequences of sex. I shall also draw attention to the evidence indicating that many of the transcending steps in the long story of life on earth have occurred by processes not encompassed by the "modern synthesis."

A VIEW OF SPECIES

Man has always been fascinated by the great diversity of organisms which live [p. 78] in the world around him. Many attempts have been made to understand the meaning of this diversity and the causes that bring it about. To many minds this problem possesses an irresistible aesthetic appeal. (Dobzhansky, 1951)

In 1952 I first read these words and became fully conscious of this irresistible appeal to the mind. This early interest in eukaryotic diversity, in turn, led to my interest in the genetic nature of species because species are the units of diversity. To address the problem of diversity from the basis of species it is evidently necessary to understand, at least in broad genetic terms, how new species might arise. Darwin failed to provide any answer which satisfies us today to the principal problem he addressed, the origin of species. It is now apparent that this is because he did not have, and indeed could not have had at that time, a clear picture of the genetic nature of species. Therefore, he was unable to address the subject of his book in an adequate way. In fact, he conceived species morphologically, though through evolutionary eyes, and set out to answer his big question from this viewpoint. Now, however, we can see that an understanding of species diversity depends on a basic understanding of the genetics of speciation, which, in turn, is constrained by our view of species.

After some twenty years of rather unquestioning support for the prevailing biological species concept (or isolation concept), during which I applied it to the study of problem species in Diptera of medical importance (e.g., Paterson, 1956, 1962, 1963, 1964a, 1964b, 1964c; Paterson and Norris, 1970), the shortcomings of defining species in terms of their reproductive isolation from other species began to become manifest. The process of disillusionment was accelerated by my reading of George Williams' (1966) book *Adaptation and Natural Selection*, which upset my understanding of natural selection and which has continued to influence me to this day. In this work Williams set out to instill some discipline and rigor into evolutionary discussions involving adaptation through natural selection. Structuralists would probably wish to avoid this approach altogether, and I sympathize with this attitude to the extent of being interested in exploring the possibility of reducing my reliance on explanations involving the concept of adaptation through the differential transmission of alternative alleles.

In searching for a more satisfactory approach to species than in terms of reproductive isolation mediated by isolating mechanisms, I went back to first principles, in this case, sex, and asked the question: How, in general terms, is the inherently improbable process of fertilization in biparental eukaryotes achieved?

A little exploration and a little thought revealed that every biparental organism must possess a set of characteristics which are effective in bringing about fertilization under the conditions prevailing in the organism's normal environments. This is really axiomatic and applies to all biparental eukaryotic unicells, to fungi, plants, and animals, including *Homo sapiens*. The characteristics of these *fertilization systems* vary with the ways of life of the organisms and with their normal environments, but in all cases known to me [p. 79] they are clearly necessary for the fundamental process of fertilization to occur. In George Williams' terms, the function of the fertilization system is the achievement of fertilization under normal natural conditions, no more and no less. The fertilization system as a whole can thus be regarded as a complex adaptation.

Before proceeding with this argument I should deal with a potent, all-pervading source of confusion. Darwin viewed species in morphological terms though he did this in quite a sophisticated way, using evolutionary explanations (Darwin, 1859:423–24). He saw the "fitness" of an organism in terms of its aptness for life in its normal environment. Accordingly, he saw the rather extreme secondary sexual characters of some species, particularly birds such as the peafowl, as leading to reduced fitness in his sense of the term. A peacock's train can hardly fit it for life in its normal wooded environment and might well reduce its chances of surviving predation. To accommodate such situations Darwin introduced the concept of *sexual selection*. Instead of simply noticing that every biparental eukaryotic organism is necessarily equipped with a system which leads effectively to fertilization, he believed that a further set of characteristics often evolved under the different selection pressure arising from either male/male competition for females, or, in some cases, interfemale competition for males. Whether sexual selection in the strict sense really is a significant factor in evolution is still far from certain, despite the burgeoning literature which assumes it is (Halliday, 1983). In any case, it is seldom noticed that if it does exist it is an additional system (complex adaptation) to the fertilization system and that it has evolved under distinct selection pressures and has a distinct function. Thus, looking at reproductive systems of organisms like peafowl or birds-of-paradise, one needs to be very cautious in disentangling the fertilization system characters from possible characters evolved in response to competition for mates. In studying particular species I believe that one needs always to ask: Can the reproductive behavior of members of this species be entirely accounted for in terms of the fertilization system alone? If the answer is yes, it is clearly unjustified to believe that they evolved in response to competition for mates. A great clarification of the literature on reproductive behavior would immediately result if this distinction were to be made consistently.

I return now to my main theme to consider a major consequence of any fertilization system appropriate to the organism's normal environment and way of life. The consequence is that such a specific fertilization system constrains the exchange of genes so that exchange effectively occurs within the group which shares the same fertilization system, the same normal habitat, and the same way of life. Obviously, such properties are the same as those characterizing any genetical species.

To sum up these points, one can define a species in genetical terms as that most inclusive population of individual biparental eukaryotic organisms which share a common fertilization system. Every fertilization system defines locally [p. 80] the field of gene recombination of a species. It is, perhaps, prudent to discuss this proposition briefly in order to provide a clear perspective. Fertilization systems can differ markedly or only to a minor degree with the consequence that there is a continuum in difference between populations. At one extreme, two populations can comprise organisms with virtually identical fertilization systems. At the other, we find organisms with fertilization systems which are wholly different from each other, as between an ostrich and a ruby-throated hummingbird. Often, as in ducks (Lorentz, 1941) or Pelecaniformes (van Tets, 1965), components of the fertilization system may be shared by different species with no loss of effectiveness.

Since my interest is in the understanding of the diversity in nature, I shall not be concerned with man-made situations. Lions and tigers do hybridize in zoos when confined to the same cage. However, the fertilization system is here destabilized because it was not evolved to be effective in a zoo cage where no appropriate mates are present. Among plants a similar situation prevails in a botanical garden where many species of related plants are artificially brought together. Another botanical example results from the destabilization of fertilization systems through the disturbance of the environment (through human or natural causes) to such a degree that populations normally found in distinct environments are brought into proximity (Anderson's 1949 "hybridization of the habitat").

A fertilization system, I have already emphasized, is appropriate for the environment in which the organism normally lives and for its way of life (whether it is sessile or motile, diurnal or nocturnal, etc.). Under these circumstances the system is very stable. The individual characters are evidently subject to stabilizing selection: aberrant individuals with inappropriate components to the fertilization system will be less likely to fertilize or be fertilized. In motile animals, particularly, an important part of the fertilization system is the specific-mate recognition system (SMRS) which serves to bring together potential mating partners. The SMRS characters are involved in the sending or receiving and processing of signals between mating partners or their cells (e.g., sperm and ovum, pollen and stigma). Sessile organisms, such as angiosperms or sessile molluscs, are much less dependent on the SMRS. Sessile organisms rely, inevitably, on vectors to transport their sex cells. These vectors are wind, water, and animals (insects, birds, and mammals). The signals between angiosperms and insects are not part of the SMRS, but are accommodations to a vector. Coadaptation between, say, a male signal and female receiver places a strong constraint on independent change in either the male or female component. The same is true for signals between a pollen grain and the stigma of the mating partner. SMRS signals, like the rest of the fertilization system, are also usually apt for the organism's normal environment. Thus fireflies, moths, and owls signal appropriately for nocturnal conditions. Forest birds and frogs are characterized by calls appropriate for transmission through dense vegetation (Morton, 1975). Birds of open [p. 81] grassland signal visually as well as by sound, as sound is readily disturbed by wind (Morton, 1975), etc.

Large population sizes of species make it nearly impossible to change a coadapted signal from one stabilized (coadapted) condition to another for the whole population. For this and other reasons, speciation is believed to involve small populations (see below), and the termination of speciation coincides with the growth of the daughter population following the switch from directional to stabilizing selection.

Thus, for three independent reasons, fertilization systems are stabilized: their relationship to the population's normal habitat, the coadaptation of signals and receivers (and other aspects of their fertilization systems and biologies), and the stability of large populations.

SPECIATION

From this view of species, one can understand a little of what must occur during the formation of a new species. The difficult task to be accomplished entails the changing of critical characters, including the fertilization system, from one stable state, appropriate to the parental environment, to a new stable state, appropriate to the new environment to which the daughter population has become restricted.

The stable state of a species in its normal habitat can be changed to a new one only if its stability is first disturbed. For example, a fertilization system will be destabilized if a small population is displaced to a distinctly different environment in which the characters of the fertilization system, etc., will not be fully effective. Stabilizing selection on these characters will then be replaced by directional selection which will lead to the spreading of alternative alleles determining a new phenotype more effective under the new conditions. A small population will facilitate the fixation of the alleles determining a new stable state. This applies to other adaptive characters such as feeding and nesting habits and predator-avoiding behavior as well. Achieving a new stable state involves the adjustment of all such characters of the organisms.

Since speciation entails small populations, the bottleneck probably facilitates the fortuitous fixation of certain alleles and chromosome arrangements at the same time. Of course, the genetic variation fueling the selective process may involve pleiotropy.

Thus, the "normal habitat" of a species is thought to approximate in key features to the one in which speciation occurred. Changing habitat is a major revolution for organisms, and available evidence convinces me that it corresponds with a speciation or subspeciation event.

In the normal habitat all environment-related characteristics of members of a species are under stabilizing selection, not just the fertilization system including the SMRS. Thus, for effective development to occur, fertilized eggs of fish, frogs, or mosquitoes, for example, must develop within certain limits. [p. 82] The development of the eggs of the subarctic mosquito, *Aedes stimulans*, normally occurs at low temperatures. If the eggs are kept at higher temperatures, development is aberrant (Horsfall and Anderson, 1963). Similarly, organisms in their normal habitats are efficient in feeding, avoiding the predators with which they are in contact, and, in some cases, nidification. This efficiency is seriously disrupted if organisms are displaced into effectively different habitats. Judging by the prevalence of simplistic models for how habitat change might occur, it is not well appreciated that change of habitat for any group of organisms requires a major genetic revolution. Mayr (1963) and others have drawn attention to the fact that speciation entails many changes in ecological characteristics, but this has not always been understood.

The work of Coope and his colleagues (see Coope, 1979, for review) on Quaternary fossil beetle assemblages in peat bogs in Europe and America has provided much telling support for these points. His demonstration of the stability of organism-habitat relationships despite the overwhelming environmental changes due to advancing or retreating ice is of fundamental importance in understanding why species are so stable and must be taken into account in formulating any views on the diversity of the living world. Coope has pointed out that speciation is not a common outcome of such catastrophic environmental events. For it to occur, organisms must be "trapped" in front of a mountain range, for example, so that retreat ahead of the change by the organisms and their environment is blocked. Even then, extinction is a more general outcome than is speciation or subspeciation.

Restriction of organisms to new environmental conditions as a necessary prelude to speciation can occur in many ways. Relic populations can commonly be observed today telling of former wider distributions under different climatic regimes. These provide us with illustrations of what Coope is saying. Increased aridity and cooling seem to be particularly effective in restricting organisms to new habitats through the progressive degrading of relic communities. Of course, islands at appropriate distances from continents provide other possible conditions which lead to speciation through destabilizing the species which invade them adventitiously. Other possibilities exist as well.

OTHER PROPOSED MODES OF SPECIATION

Many other modes of speciation have been proposed (White, 1978; Mayr, 1963, 1987), but they have all been conceived with the constraints of the isolation concept in mind. It should therefore be noted that the model of speciation which I call allopatric speciation is genetically different from the well-known model of geographic or peripatric speciation long advocated by Ernst Mayr (1942, 1954).

Mayr's model and mine share common ground in requiring the physical isolation of small populations for sufficient time for speciation to occur but differ because Mayr was obliged by his species concept to rely on pleiotropy to [p. 83] modify the existing "primary" isolating mechanism (usually some form of postmating isolating mechanism). He believed that other isolating mechanisms often evolved later under natural selection when daughter and parent populations again met (Mayr, 1963:551). I have previously provided details of the difficulties existing for the various models generated by the isolation concept (Paterson, 1978 [chap. 3], 1982a [chap. 8], 1985 [chap. 12]) and have argued that no prima facie evidence exists to show that any species has ever arisen by any model of speciation other than some form of allopatric speciation (Paterson, 1981 [chap. 5], 1985 [chap. 12], 1988 [chap. 14]) and that substantial difficulties confront models invoking selection for reproductive isolation (Paterson, 1978 [chap. 3], 1985 [chap. 12]).

It should be reiterated that models of speciation are logically derived from concepts of species so that when a concept of species is abandoned, all its dependent speciation models must be discarded with it.

CONSEQUENCES

Interesting genetic and ecological consequences flow from the picture of species in terms of specific-mate recognition and how they are likely to arise. These consequences lead us to look at the data of nature with different eyes. The prevailing concept of species, the isolation concept of Dobzhansky and Mayr, defines one species in relation to other species, which leads to many problems which I have outlined before (Paterson, 1978 [chap. 3], 1980 [chap. 4], 1981 [chap. 5], 1982a [chap. 8], 1985 [chap. 12], 1988 [chap. 14]). Derived consequences from the isolation concept are quite different from those from the recognition concept, despite Mayr's (1987) assertion that the two concepts can be seen as "the two sides of a single coin." It is not possible to compare the consequences of the two concepts here, but some consequences of the recognition concept will be discussed.

Although natural selection is invoked as part of the process of speciation, it is merely part of the process of organisms adapting to new conditions. If speciation results from this it is fortuitous and is an extreme of a continuum in degrees of divergence. In George Williams' terms, species are not "adaptive devices" but incidental consequences of adaptation ("effects"). This is a fundamentally important insight. Darwin reached a similar conclusion but with a limited understanding of species and speciation. To many this conclusion is unacceptable when applied to our own species (Paterson, 1982b [chap. 9], 1985 [chap. 12], 1988 [chap. 14]). However, it has a far wider significance than this. For example, it affects fundamentally our views on ecology and the causes of biological diversity (Paterson, 1985 [chap. 12], 1986 [chap. 13]; Walter et al., 1984; Hulley et al., 1988).

The model of speciation which follows from the recognition concept also directs thinking in particular directions and has important implications. It very firmly invokes speciation in small populations, despite population genetical reservations on the subject. This is in keeping with empirical evidence from actual speciation events of the past, and it provides a better understanding of why "punctuation" should be a feature of the fossil record (Eldredge, 1971; Eldredge and Gould, 1972). The recognition concept is also [p. 84] important in understanding the assembly of ecosystems and communities and in interpreting biogeographical data.

The three reasons for the stability of species provide insights for the understanding of a number of aspects of evolution. They provide a clear explanation of the "equilibrium" phase of Eldredge and Gould's "punctuated equilibria" view of the fossil record. Species homeostasis also has an important bearing on the long and drawn out debate on sympatric speciation. Member organisms of a species in their normal habitats are subject to mainly stabilizing selection. There is no pressure to speciate under such conditions because there is nothing to disturb the existing stable state. This view contrasts with the view of species as "adaptive devices," which diverge in order to exploit more fully natural resources by occupying "unoccupied niches." Under the recognition concept this viewpoint is wholly unjustified. The niche occupied by the organisms of a species is not a property of the environment but is largely a reflection of their genotype (Paterson, 1973; Walter et al., 1984). Once this is understood it is no longer a puzzle why species do not occupy "niches" which appear to ecologists as so temptingly "vacant." Expectations are likely to remain unfulfilled if they stem from inappropriate analogies with human economic systems.

When species are understood to be extreme incidental consequences of adaptation to new environments, hybrids and hybridization are seen in quite new ways. As J. L. Crosby (1970) once warned, the consequences of hybridization may be good or bad and are decided by natural selection, not by our biases and idealistic preconceptions. Once the preservation of "species integrity" or "species purity" is seen to involve deep cultural preconceptions (Paterson, 1982b [chap. 9], 1985 [chap. 12], 1988 [chap. 14]) we stop expecting hypothetical "isolating mechanisms" to be "reinforced" by selection and expect natural selection simply to act to maintain viability. A full understanding of what is occurring in hybrid zones or in hybrid populations demands a proper understanding of the nature of species.

Species integrity is conventionally seen as being protected from destruction by so-called isolating mechanisms. Few pause to wonder why these isolating mechanisms are often protecting hybrid genomes, as in the case of polyploid species of plants. These are widely accepted as generally being allopolyploids (amphiploids). Are the isolating mechanisms in such cases still protecting "species integrity"?

Some authors (Scoble, 1985; Templeton, 1987) criticize the recognition and isolation concepts for not covering uniparental organisms. Both Dobzhansky and I have accepted that this desire for generality has its roots in taxonomy (see Mayr's 1963:28 approach), but reject it because our aim is to understand natural diversity, not to classify organisms. Most eukaryotes are biparental, or are at least facultatively biparental, and so are to be understood in terms of a genetical concept. A few primary uniparental eukaryotic organisms may exist (amoebae, euglenoids), but most are derived from biparental organisms. [p. 85] Widening a species concept in order to cover these secondarily uniparental organisms destroys the explanatory power of a genetical concept and involves the conflation of concepts of species from two quite distinct fields of scientific endeavor (taxonomy and evolutionary genetics) (Paterson, 1981 [chap. 5]). Rather than attempting to devise a concept that brings "chalk and cheese" under one heading, I prefer to account for the secondarily uniparental eukaryotes by trying to understand their origin. Understanding that they very generally bear signs of a hybrid origin, uniparental eukaryotes (e.g., allopolyploid plants) can often be recognized as examples of organisms which have "escaped from hybridity" (Darlington, 1958; Grant, 1971; Paterson, 1981 [chap. 5], 1985 [chap. 12],

1988 [chap. 14]). This alternative viewpoint is in accordance with the idea that, as usual, natural selection acts on hybrids by selecting for viability (Paterson, 1978 [chap. 3]).

This brief catalog is not comprehensive in listing the important new insights deriving from the adopting of the intuitively appealing, and evidently effective, recognition concept of species. Active application of the ideas will readily reveal many more.

DISCUSSION

> The thesis that new facts are always responsible for the ultimate clarification of scientific problems is becoming increasingly questionable. (Mayr, 1976:331)

Mayr is here drawing attention to the fact that the way we look at problems in science is often more critical in clarifying them than is the provision of yet more empirical data.

Some evolutionists still feel happy with the "evolutionary synthesis." This is depressing to me and, I am sure, to many others, in the face of the remarkable empirical and theoretical advances of this period. In fact, I believe the advances made in the last forty years have revolutionized our evolutionary view of life, and what is new should be reflected in our synthesis. On the other hand, it would be surprising if nothing remained of the very impressive edifice erected in the thirties and forties. Perhaps reformatting our views on evolution should be our aim rather than debating whether a new synthesis is required or not.

To many, the core of the evolutionary synthesis is the biological species concept. Certainly much in the synthesis is related to species theory. It follows, therefore, that a significant change in viewpoint on the genetic nature of species is likely to be an important reason to reformulate the evolutionary synthesis. As the outline provided above will have demonstrated, the recognition concept does constitute a new way of looking at species in genetic terms, and its adoption will thus ensure that any evolutionary synthesis built around it will be significantly new.

Besides, we must also incorporate the appreciation that so transcending a step as the evolution of the eukaryotic cell entailed a process not envisaged under the world view of the evolutionary synthesis: symbiosis. Today we [p. 86] appreciate that symbiosis involves a radical form of genetic and systems recombination quite unimagined or, at least, unconsidered in 1942. The evolutionary synthesis has never provided a satisfactory explanation of the evolution of sex. The cannibalism theory of the evolution of sex (Margulis and Sagan, 1986) provides a reasonable scenario of how sex might have evolved. This model cannot be considered part of the evolutionary synthesis despite its having been considered seriously by Maynard Smith in 1958. We might eventually achieve a more credible and general understanding of the evolution of sociality when we rethink it in comparable stochastic terms and abandon the procrustean task of forcing the problem into a neo-Darwinist framework. The neutral theory of molecular evolution (King and Jukes, 1969; Kimura, 1968) has provided many insights which would not have been forthcoming in the light of the evolutionary synthesis, and I am sure there are more insights to come from it.

Although I have not exhausted the revolutionary ideas outside the evolutionary synthesis, enough examples of major consequence have been cited to illustrate the fact that we now have at our disposal the elements of a quite radically different view of evolution than was available in the thirties and forties. A synthesis of these elements has not yet been made, because, besides its being a formidable task for which few are fitted, I believe few have attempted to think through the consequences. Or perhaps those who have done so have drawn back in awe at the picture partially glimpsed.

Much of this modified world view will be acceptable to structuralists, but probably not all. At present I still retain a place for functionalist explanations, though I check their credentials individually. But I certainly do not exclude the explanations of physical scientists or structuralists (Rosenberg, 1985). I don't believe this stand constitutes fence-sitting, as that has not really been my style in the past; rather I think it is because both kinds of explanation can, at times, provide understanding. A hope we might cherish in support of Mayr's statement with which we began this discussion is that scientists will in future spend more time in considering theory instead of generating more and more empirical data in attempting to answer inappropriately posed questions.

REFERENCES

Anderson, E. 1949. *Introgressive Hybridisation*. New York: John Wiley.

Coope, G. R. 1979. Late Cenozoic fossil Coleoptera: Evolution, biogeography, and ecology. *Annual Review of Ecology and Systematics* 10:247–67.

Crosby, J. L. 1970. The evolution of genetic discontinuity: Computer models of the selection of barriers to interbreeding between species. *Heredity* 25:253–97.

Darlington, C. D. 1958. *Evolution of Genetic Systems*. 2d ed. Edinburgh: Oliver and Boyd.

Darwin, C. 1859. *On the Origin of Species by Means of Natural Selection; or, the Preservation of Favoured Races in the Struggle for Life*. London: John Murray.

Dobzhansky, T. 1951. *Genetics and the Origin of Species*. 3d ed. New York: Columbia University Press.

Eldredge, N. 1971. The allopatric model and phylogeny in Paleozoic invertebrates. *Evolution* 25:156–67.

Eldredge, N., and S. J. Gould. 1972. Punctuated Equilibria: An Alternative to Phyletic Gradualism. Chap. 5 in *Models in Paleobiology*, ed. T.J.M. Schopf, 82–115. San Francisco: Freeman, Cooper and Company.

Grant, V. 1971. *Plant Speciation*. New York: Columbia University Press.

Halliday, T. R. 1983. The Study of Mate Choice. In *Mate Choice*, ed. P. Bateson, 3–32. Cambridge: Cambridge University Press.

Horsfall, W. R., and J. F. Anderson. 1963. Thermally induced genital appendages on mosquitoes. *Science* 141:1183–84.

Hulley, P. E., G. H. Walter, and A.J.F.K. Craig. 1988. Interspecific competition and community structure, I. *Rivista di Biologia–Biology Forum* 81: 57–71.

Kimura, M. 1968. Evolutionary rate at the molecular level. *Nature* 217: 624–26.

King, J. L., and T. H. Jukes. 1969. Non-Darwinian evolution: Random fixation of selectively neutral mutations. *Science* 164:788–98.

Lorentz, K. 1941. Vergleichende Bewegungsstudien an Anatinen. *Journal für Ornithologie (Suppl.)* 89:194–294.

Margulis, L., and D. Sagan. 1986. *Origins of Sex*. New Haven, Conn.: Yale University Press.

Maynard Smith, J. 1958. *The Theory of Evolution*. Harmondsworth, U.K.: Penguin Books.

Mayr, E. 1942. *Systematics and the Origin of Species*. New York: Columbia University Press.

———. 1954. Change of Genetic Environment and Evolution. In *Evolution as a Process*, ed. J. S. Huxley, A. C. Hardy, and E. B. Ford, 157–80. London: Allen and Unwin.

———. 1963. *Animal Species and Evolution*. Cambridge, Mass.: Belknap Press of Harvard University Press.

———. 1976. Sibling or Cryptic Species among Animals. In *Evolution and the Diversity of Life*, ed. E. Mayr, 509–14. Cambridge, Mass.: Harvard University Press.

———. 1987. The Species as Category, Taxon and Population. Chap. 15 in *Histoire du Concept d'Espèce dans Les Sciences de la Vie*, ed. J. Roger and J.-L. Fischer, 303–20. Colloque international (mai 1985) organisé par la Fondation Singer-Polignac, Paris. Paris: Fondation Singer-Polignac.

Morton, E. S. 1975. Ecological sources of selection on avian sounds. *American Naturalist* 109:17–34.

Paterson, H.E.H. 1956. Status of the two forms of housefly occurring in South Africa. *Nature* 178:928–29.

———. 1962. Status of the East African salt-water-breeding variant of *Anopheles gambiae* Giles. *Nature* 195:469–70.

———. 1963. The species, species control and antimalarial spraying campaigns: Implications of recent work on the *Anopheles gambiae* complex.

South African Journal of Medical Science 28:33–44.

———. 1964a. Saltwater *Anopheles gambiae* on Mauritius. *Bulletin of the World Health Organisation* 31:635–44.

———. 1964b. Population Genetic Studies in Areas of Overlap of Two Subspecies of *Musca domestica* L. In *Ecological Studies in Southern Africa*, ed. D.H.S. Davis, 244–54. The Hague: W. Junk.

———. 1964c. Direct evidence for the specific distinctness of forms A, B, and C of the *Anopheles gambiae* complex. *Rivista di Malariologia* 43:191–96.

———. 1973. Animal species studies. Pp. 31–36 in Animal and plant speciation studies in Western Australia, by H.E.H. Paterson and S. H. James. *Journal of the Royal Society of Western Australia* 56:31–43. [See this volume, Author's Preface.]

———. 1978. More evidence against speciation by reinforcement. *South African Journal of Science* 74:369–71. [This volume, chap. 3.]

———. 1980. A comment on "Mate Recognition Systems." *Evolution* 34:330–31. [This volume, chap. 4.]

———. 1981. The continuing search for the unknown and unknowable: A critique of contemporary ideas on speciation. *South African Journal of Science* 77:113–19. [This volume, chap. 5.]

———. 1982a. Perspective on speciation by reinforcement. *South African Journal of Science* 78:53–57. [This volume, chap. 8.]

———. 1982b. Darwin and the origin of species. *South African Journal of Science* 78:272–75. [This volume, chap. 9.]

———. 1985. The Recognition Concept of Species. In *Species and Speciation*, ed. E. S. Vrba, 21–29. Transvaal Museum Monograph No. 4. Pretoria: Transvaal Museum. [This volume, chap. 12.]

———. 1986. Environment and species. *South African Journal of Science* 82:62–65. [This volume, chap. 13.]

———. 1988. On defining species in terms of sterility: Problems and alternatives. *Pacific Science* 42:65–71. [This volume, chap. 14.]

Paterson, H.E.H., and K. R. Norris. 1970. The *Musca sorbens* complex: The relative status of the Australian and two African populations. *Australian Journal of Zoology* 18:231–45.

Rosenberg, A. 1985. *The Structure of Biological Science*. Cambridge: Cambridge University Press.

Scoble, M. J. 1985. The Species in Systematics. In *Species and Speciation*, ed. E. S. Vrba, 31–34. Transvaal Museum Monograph No. 4. Pretoria: Transvaal Museum.

Templeton, A. R. 1987. Species and speciation. *Evolution* 41:233–35.

Tets, G. F. van. 1965. *A Comparative Study of Some Social Communication Patterns in the Pelecaniformes*. Ornithological Monographs No. 2: American Ornithologists' Union.

Walter, G. H., P. E. Hulley, and A.J.F.K. Craig. 1984. Speciation, adaptation and interspecific competition. *Oikos* 43:246–48.

White, M.J.D. 1978. *Modes of Speciation*. San Francisco: W. H. Freeman.

Williams, G. C. 1966. *Adaptation and Natural Selection*. Princeton, N.J.: Princeton University Press.

16 Updating the Evolutionary Synthesis

Paterson, H.E.H. 1989. Updating the evolutionary synthesis. *Rivista di Biologia—Biology Forum* 82:371–75.

I presented this paper at the International Symposium on Fundamental Problems in Evolutionary Biology in Moscow in April 1989. Many delegates were committed structuralists, and so the meeting was really the Third International Workshop of Structuralism in Biology. My subject was the current state of health of the evolutionary synthesis. Against the trend to regard the synthesis as dead, I argued that we should acknowledge that one inevitably builds on the past and that it is unrealistic to attempt to break up the continuum in some artificial way. I pointed out some glaring shortcomings of the modern synthesis and urged that we update it by building on what is sound.

Although the evolution of sex has, since the early 1970s, been a central topic among those of us who advocate the recognition concept of species, we have published little on our position. This paper includes an indication of my thoughts on the subject. I also use the term "macrorecombination" for hybridization and symbiosis, where independent genomes which have often been separately evolving for millions of years are recombined. Sometimes the consequences are of transcending significance, as with the evolution of the eukaryotic cell, the evolution of sex, and many cases of speciation via hybridization. Such events show up the problems facing those who approach evolution through phylogenetic systematics.

I speak of the "evolutionary synthesis" because I am unhappy with the term "neo-Darwinism" which has been applied over many years to a number of rather distinct movements. In any case, terms ending in "ism" need cautious treatment because they are generally compounds, not simples, which makes them slippery to deal with. The "evolutionary synthesis" is, in any case, more palpable because its history has been recorded (Mayr and Provine, 1980).

Revising the evolutionary synthesis requires a number of methodological improvements on past practices. For example, authors need to be more consistent than have been the founders of the movement. Past inconsistencies have led to much unnecessary misunderstanding and disputation. Conflation of concepts has been widespread, and so has a tendency to misread other authors through the preconception of their messages. Above all, a broad vision is needed to encompass both the detail of modern analytical studies and the importance of broader theoretical works.

Why I believe that an updating of the evolutionary synthesis is called for might best be made apparent by identifying a number of "transcending steps" (Dobzhansky and Ayala, 1977) in the long course of evolution, and then considering how they are treated by the authors of the synthesis and by others.

Toward this end, I shall in this brief presentation consider three such transcending steps and their consequences. The origins of the eukaryotic cell, of mitosis, and of sex have all been transcending in Dobzhansky's sense, and all are difficult to account for under the modern synthesis and generally receive cursory or no attention from authors. [p. 372]

A great advance occurred with Margulis's (1970) revival and rejuvenation of an old idea which attributed a major role to endosymbiosis in accounting for the origin of eukaryotic cells. By this process the genome of an anoxic prokaryote was combined with the genome of an oxygen-respiring one. This hypothesis has subsequently been subjected to close scrutiny and extensive testing and is consequently now widely subscribed to. Photosynthetic eukaryotes were similarly contingent upon the endosymbiotic incorporation of a third genome from a photosynthetic prokaryote. Margulis has further proposed that unicellular eukaryotes have acquired motility by the endosymbiotic incorporation of motile, spirochaetelike cells in much the same manner as can be found in the protistan *Myxotricha paradoxa* described by Grimstone and Cleveland (1964). Although less secure than other aspects of Margulis's model, this suggestion may yet receive adequate support.

The evident similarities in structure between the basal bodies of cilia, flagella, and centrioles, and their relationship to the microtubules of the mitotic spindle, suggest that the acquisition of such spirochaetelike endosymbionts was the initiating step in the origin of mitosis. Mitosis, in turn, was evidently an obligatory step on the path to the acquisition of meiosis and sex by eukaryotes (Margulis and Sagan, 1986).

Among eukaryotic organisms sex can be characterized as the alternation of meiosis, fertilization, and syngamy. Accounting for the origin of so complex a system as sex among early eukaryotic unicells by the sequential addition of steps, as expected under the synthesis, poses insuperable difficulties. The origin of the fertilization system might be understandable in this way once meiosis and syngamy had evolved, but not before. Similarly, meiosis and syngamy have significance only in the complete system. Complex characters for which a role can be recognized seem sometimes to stem from fortuitous events. Once the general configuration of a complex character is arrived at through a stochastic event, its effectiveness can subsequently be improved by conventional selection. In accounting for the origin of sex, authors (e.g., Maynard Smith, 1958; Margulis and Sagan, 1986) have suggested that cannibalism by a eukaryotic unicell might have provided the fortuitous start to the process. The act of cannibalism here amounts to fertilization, and the combining of the two haploid nuclei to form a diploid cell would be a primitive form of syngamy. The subsequent mitotic event could conceivably result in recombination by the independent assortment of chromosomes into the daughter haploid cells not dissimilar to the "one-stage" meiosis among [p. 373] present-day protista (Cleveland, 1947; Margulis and Sagan, 1986) or even a "two-stage" meiosis (Sleigh, 1973:79-80).

I shall not consider here the selective basis for the improvement of this primitive meiosis under selection. Some (Maynard Smith, 1958; Margulis and Sagan, 1986; Michod and Levin, 1987) have seen its advantage in terms of genetic repair, but the direct advantages of recombination may prove to be the basis of the selective process. However, an essential early addition to the evolving system of sex must have been the acquisition of a recognition system which facilitated the achievement of the cell fusion by like cells, which at first constituted "fertilization" (Maynard Smith, 1958; Paterson, 1973; Margulis and Sagan, 1986). When one cell responds to a (chemical) signal emitted by another cell, we can refer to the process as "recognition." The process is comparable to the recognition of a specific antigen by a particular antibody. The advent of biparental organisms constituted the first example of a new biological unit, the genetical species, the unit of eukaryotic diversity. A genetical species is defined by me as *that most inclusive group of biparental organisms which shares a common fertilization system* (Paterson, 1985 [chap. 12]). I have called this way of viewing species the "recognition concept of species." The acquisition of recombination freed eukaryotes from their early dependence on mutation as the main basis for change and the tyranny of a clonal population structure (Miller and Hartl, 1986).

With the first species came the potential for the indefinite initiation of daughter species (Paterson, 1985 [chap. 12]) through the local fixation of genetically determined changes to the original fertilization system. Thus was commenced the process that underlies the present diversity of life on earth. To list the consequences of the attainment of the eukaryotic state, mitosis, and sex is a formidable task. However, some impression can be obtained by considering the structural and functional diversity of eukaryotic organisms and contrasting this picture with that gained by considering the prokaryotes. These developments all depend on the prior acquisition of the three steps which I have identified here, and all depend on instances of radical recombination resulting from the establishment of serial endosymbiotic associations. These associations bring together in one organism millions of years of divergent, independent evolution. One would search in vain in the major texts of the evolutionary synthesis for these insights. The index of the historical account of the synthesis (Mayr and Provine, 1980) does not contain the word "symbiosis," nor a reference to Lynn Margulis. In Mayr's monumental work, *The Growth of Biological* [p. 374] *Thought* (Mayr, 1982), Margulis is referred to very briefly on two occasions. Futuyma (1986) made only brief reference to these insights, not giving any impression of their transcending nature.

Along the complex chain of consequences of the founding of the eukaryotic cell, then mitosis, sex, and species, we can detect another form of radical recombination which has also played an important role in evolution but which has not received the attention it warrants. I refer here to speciation following hybridization. Among angiosperms, at least, many species have a hybrid origin (Grant, 1963:487). In each such case, two distinct genotypes, each the product of long divergent adaptation to distinct environmental conditions, are combined in a single organism. I suggest extending Grant's (1963:470) term "macrorecombination" to cover not only this type of recombination, but also the recombination events that led to eukaryotic cells, photosynthetic eukaryotes, and mitosis. These points have been brought together here to illustrate the fact that some of the most crucial events along the path of evolution are scarcely noticed or attended to under the evolutionary synthesis, which supports my contention that it is in need of serious updating. I find unproductive the type of defense of the status quo which simply absorbs all objections to the synthesis by denying that they are, in fact, objections (Mayr, 1985). It is surely time to cease shoring up the currently inadequate synthesis and to begin its reconstruction.

REFERENCES

Cleveland, L. R. 1947. The origin and evolution of meiosis. *Science* 105:287–88.

Dobzhansky, T., and F. J. Ayala. 1977. *Humankind: A Product of Evolutionary Transcendence*. Johannesburg: Witwatersrand University Press.

Futuyma, D. J. 1986. *Evolutionary Biology*. Sunderland, Mass.: Sinauer Associates.

Grant, V. 1963. *The Origin of Adaptations*. New York: Columbia University Press.

Grimstone, A. V., and L. R. Cleveland. 1964. The structure of *Myxotricha* and its associated microorganisms. *Proceedings of the Royal Society (B)* 159:668–86.

Margulis, L. 1970. *Origin of Eukaryotic Cells*. New Haven, Conn.: Yale University Press.

Margulis, L., and D. Sagan. 1986. *Origins of Sex*. New Haven, Conn.: Yale University Press.

Maynard Smith, J. 1958. *The Theory of Evolution*. Harmondsworth, U.K.: Penguin Books.

Mayr, E. 1982. *The Growth of Biological Thought*. Cambridge, Mass.: Harvard University Press.

———. 1985. What is Darwinism today? *Philosophy of Science Association* 1984:145–56.

Mayr, E., and W. B. Provine. 1980. *The Evolutionary Synthesis*. Cambridge, Mass.: Harvard University Press.

Michod, R. E., and B. R. Levin. 1987. *The Evolution of Sex*. Sunderland, Mass.: Sinauer.

Miller, R. D., and D. L. Hartl. 1986. Biotyping confirms a nearly clonal population structure in *Escherichia coli*. *Evolution* 40:1–12.

Paterson, H.E.H. 1973. Animal species studies. Pp. 31–36 in Animal and plant speciation studies in Western Australia, by H.E.H. Paterson and S. H. James. *Journal of the Royal Society of Western Australia* 56:31–43. [See this volume, Author's Preface.]

———. 1985. The Recognition Concept of Species. In *Species and Speciation*, ed. E. S. Vrba, 21–29. Transvaal Museum Monograph No. 4. Pretoria: Transvaal Museum. [This volume, chap. 12.]

Sleigh, M. A. 1973. *The Biology of Protozoa*. London: Edward Arnold.

17 The Recognition of Cryptic Species among Economically Important Insects

Paterson, H.E.H. 1991. The Recognition of Cryptic Species among Economically Important Insects. Chap. 1 in Heliothis: *Research Methods and Prospects*, ed. M. P. Zalucki, 1–10. New York: Springer-Verlag.

Although this paper is aimed at applied entomologists, it is the only paper I have published on sibling species. This is curious because, over my research career, I have specialized in seeking to understand complexes of cryptic species. This is also true of many of my former students. As a group we have detected sibling species in Lepidoptera, Anura, Rodentia, Crustacea, Diptera, Primates, and even in that best worked group, the birds.

The recognition concept provides for the first time a clear understanding of the nature of sibling species and discredits the common idea that sibling species are recently formed species which will become well-marked species with time (Paterson, 1982, p. 275 [chap. 9]).

I believe that a major consequence of our studies on practical and theoretical aspects of sibling species is to make ecologists, including economic and medical entomologists, aware of the flawed nature of much work that ignores the problem of recognizing the limits of gene pools. It is a fact that vast sums of money are wasted through ignoring this fundamental difficulty. But, worse still, much ecological work is confused and rendered useless through inadequate attention to the fundamental need to distinguish the limits of gene exchange in the material under study.

INTRODUCTION

Medical and economic entomologists are concerned with understanding the ecology of particular insect pest species, that is, their abundance and distribution. The fundamental problem that needs to be confronted is that we cannot always recognize the species we are studying. This is because of the common occurrence of cryptic species that cannot be separated by traditional taxonomic methods, no matter how carefully they are applied. For this reason, a species taxon defined purely on morphological criteria does not coincide necessarily with a species delimited by evolutionary geneticists. This impediment to scientific study can be overcome by admitting the problem and then dealing with it by bringing to bear existing genetical techniques guided by appropriate evolutionary genetical insights.

In this chapter, I briefly outline the difference between a morphological and a genetical species as well as those methods available to detect cryptic species. I also discuss some examples pertinent to (Australian) agricultural practice, with some emphasis on the species of *Heliothis*.

Central to my theme is the fact that the problem will not be solved by techniques alone, no matter how up-to-date they may be. The essential need is to be able to frame critical questions, informed by evolutionary insights. Only then do techniques become useful and important. It is often forgotten that techniques are tools that can be used by workers with skill and imagination, and what results from their use depends entirely on the conceptual grasp of the user. With this in mind, I have taken trouble to outline [p. 2] briefly some of the ideas needed to use available genetical tools effectively (see Daly, 1991).

KINDS OF SPECIES

The term "species" has a number of quite different meanings, a fact which is not always well understood by biologists. As a result of this, we often do not know what kind of species is under discussion. For example, much confusion results when the word "species" is used to mean morphological species by a speaker, but it is understood by a listener to mean one defined on genetic grounds, or vice versa.

In taxonomy, we deal with the classification and naming of organisms. Traditionally, we have relied on taxonomists to identify and name our specimens. To the general benefit of biology, taxonomists have amassed our great collections and have ordered them according to the system of hierarchical categories of Linnaeus. In at least 99 percent of cases, assignment of a specimen to any category from species to kingdom is done by interpreting distinct and shared structural characters. Taxonomic classification based on structure is equally effective in dealing with biparental (sexual) organisms and uniparental (parthenogenetic or asexual) ones. For example, taxonomy handles the uniparental species of morabine grasshopper *Warramaba virgo* in the same manner as it deals with its closely related biparental congeners (Key, 1976:64–66).

On the other hand, evolution is a process involving genes in populations and is studied appropriately through the insights of genetics. In evolutionary genetics, which amounts to the population genetics of natural populations, species are necessarily recognized on genetical criteria. Fortunately, the genetical species often coincides with the morphological species, although not always. Quite commonly, a species taxon comprises two or more undetected cryptic (sibling) genetical species. These would not have been missed by the taxonomist through incompetence, but rather because the cryptic species concerned often really do look identical. If one relies on visual morphological criteria to erect a species, the conclusion is inevitable that specimens that look identical are part of a single species. The nature of cryptic species will become clear after genetical species have been discussed.

There is general agreement among evolutionary geneticists that a genetical species can be regarded as "a field for gene recombination" (Carson, 1957). This simply means that members of a genetical species can exchange and recombine genes freely with other members of the same species, but generally not with members of other species. Debate still occurs over what restricts the exchange of genes to this "field," and two explanations are competing currently.

Most evolutionary geneticists believe this is because of a class of adaptations, [p. 3] the isolating mechanisms, as was suggested originally by Dobzhansky in 1935 and supported subsequently by Mayr (1963). I have abandoned this "isolation concept" (Paterson, 1978 [chap. 3], 1985 [chap. 12]) because there is little or no evidence that such characters have evolved to "protect the integrity of the genetic system of species" as supporters of the view claim (Mayr, 1963:109), and because of many internal inconsistencies with the concept (Paterson, 1985 [chap. 12]).

Because biparental organisms in their normal habitats generally exchange genes only with other members of the same species, there is good evidence that they always possess mechanisms to attract and recognize mating partners of the appropriate kind as a prelude to achieving successful fertilization. In other words, a species is that most inclusive group of biparental organisms that shares a common fertilization system (Paterson, 1982 [chap. 8], 1985 [chap. 12]).

By fertilization system, I mean all those characteristics that contribute to achieving fertilization under conditions that generally prevail in the organism's normal habitat. In motile organisms such as insects, the fertilization system always includes a signaling system involving the potential mating partners or their sex cells. I call this signaling system the specific-mate recognition system, or SMRS. It is the transmission of signals, their reception, and processing that enables males and females of an insect species to detect and "recognize" each other as a prelude to mating. After potential mating partners have come together, there are further exchanges of signals that are also necessary to achieve fertilization. Ultimately, egg and sperm "recognize" each other chemically. By recognize I mean recognition of the sort involved in the recognition of an antigen by an antibody; I do not wish to imply choice of any sort. The SMRS includes reproductive behavior or courtship, but is more extensive than that as it includes the exchange of signals by egg and sperm or pollen and stigma.

This way of understanding species in terms of mate recognition is a common sense one, which seems to apply perfectly well to all biparental, sexual organisms from protozoa to *Homo sapiens*. It is very clear, too, that, in their normal habitats, males and females of cryptic species, which look identical to taxonomists, find and recognize each other and mate without difficulty. Because they must be using their fertilization system to do this, it seems logical to recognize that it is this that sets the limits to the species gene pool or "field of gene recombination." Indeed, what could be more fundamental and sensible than for us to recognize species by the same system that the organisms themselves use when finding a mate?

Supporters of the isolation concept have attempted to explain cryptic species in a variety of ways (Mayr, 1942, 1963; Dobzhansky, 1951; Sokal, 1973; Ayala et al., 1974), but this has proved difficult in terms of reproductive isolation. The recognition concept of species, in contrast, elucidates the nature of cryptic species. Cryptic or sibling species, in fact, are not species that have yet to diverge more fully. They are [p. 4] normal genetical species in every way and cause difficulty only because human taxonomists use their optical sense in identifying species. If humans were as well provided for with other senses as insects are, species would not be cryptic. Another point that should be remembered is that it has been well established (for example, by Lambert and Paterson, 1982 [chap. 7]) that morphological resemblance is not always a reliable indication of phylogenetic closeness. Morphologically very different species are sometimes much more closely related than two cryptic species may be (for example, *Drosophila silvestris* and *D. heteroneura* [Johnson et al., 1975]). Because cryptic or sibling species use mostly nonvisual signals in their specific-mate recognition systems, we can expect cryptic species to occur most frequently in groups that live under conditions where visual signaling is inefficient or where visual signals are transient and cannot be preserved in a museum specimen. This point is well made by comparing diurnal Lepidoptera with nocturnal ones. Butterflies are diurnal and rely mainly on visual signals in finding mates (Tinbergen et al., 1942). Similarly, there are diurnal moths that utilize optical signals in their SMRSs, and these are colorful. By and large, there is a good level of coincidence between taxonomic species and genetical species in such groups. The difficulties arise in such groups as the Noctuidae and Geometridae, which are nocturnal and, therefore, rely largely on chemical and tactile signals. It is in such groups that cryptic species are common. In fireflies, on the other hand, visual signals occur though these insects are nocturnal. Of course, the signals are transient and of little use in general museum taxonomy. In contrast, to the evolutionary geneticist they are critically important, as they are to the organisms themselves. To sum up: cryptic species are likely to be found in any group of organisms that use mainly auditory, chemical, tactile, or transient optical signals in the critical, discriminating stages of their SMRS.

PRACTICAL METHODS FOR DETECTING CRYPTIC SPECIES

The existence of cryptic species is, ipso facto, often unsuspected. First, I discuss ways by which they have been disclosed most often in the past.

Most commonly, one comes to suspect the existence of a species complex from the occurrence of biological discontinuities, either local or geographical. For example, the genetical species *Anopheles melas* and *An. merus*, two cryptic species, were detected within the species taxon *Anopheles gambiae* through their ability to breed in highly saline waters in West and East Africa, respectively. This is quite unlike other populations of the species taxon, which breed only in freshwater (Muirhead-Thomson, 1948, 1951; Paterson, 1962) or, at most, in mineral waters. The species taxon *Perthida glyphopa* occurs as a leaf-mining parasite of jarrah (*Eucalyptus marginata*), [p. 5] and Common's type material was limited carefully to specimens from this host. However, morphologically very similar specimens are found on flooded gum (*E. rudis*) and on prickly bark (*E. todtiana*). Using electrophoretic and distributional studies, Mahon et al. (1982) provided evidence for the view that distinct genetic species occur on jarrah and flooded gum and that the specimens on prickly bark may form also a distinct, third, gene pool.

With the increasing interest of applied biologists in the use of sex pheromones in the biological control of insect pests, it is not surprising that evidence for cryptic genetical species has emerged fortuitously (Cardé et al., 1978), though such evidence is not always correctly interpreted by the finders.

Interesting work by Whittle et al. (1987) can be cited to illustrate how sibling species may be detected through biological clues. A tortricid moth species, *Homona spargotis*, a pest of avocados in northern Queensland, was synonymized with a Sri Lankan coffee pest, *H. coffearia*, on morphological grounds. However, pheromone analysis revealed significant differences in this important constituent of specific-mate recognition systems between the Sri Lankan and Australian moths. This led to the discovery of small but consistent morphological differences.

While studying the inheritance of dieldrin resistance in a northern Nigerian population of the species taxon *Anopheles gambiae s. str.*, Davidson (1956, 1958) crossed it with a susceptible strain from Lagos. Incidentally, he found that the F_1 males were sterile, but attributed this to a pleiotropic effect of the resistance allele. In the latter paper, he mistakenly rejected the possibility that two species were involved, as was suggested by Holstein (1957), in favor of his original conclusion. Much later it became evident that the two strains were from distinct genetical species, respectively, *An. arabiensis* and *An. gambiae, s. str.*, and the sterility observed was hybrid dysgenesis. The correct interpretation of evidence depends on a detailed understanding of the genetics of species. In widespread species, differences in such non-reproductive characters as the ability to enter diapause may also provide pointers to the existence of cryptic species.

Evidence of broad polyphagy in a species taxon might indicate that a complex of cryptic genetical species is involved. In other words, each cryptic species might account for part of the wide range of hosts exploited by the species taxon concerned. For this reason, the possible existence of cryptic species should be investigated in any critical study of the host range of an apparently polyphagous species, particularly when the species taxon is cosmopolitan. It is by no means clear, for example, that populations of *H. armigera* around the world represent only one genetical species. Preliminary studies (Daly and Gregg, 1985) have not altogether settled this problem.

It is naive to cite changes of host relationships as evidence for the adaptability of an insect species to new hosts, unless it has first been demonstrated [p. 6] that only one genetical species is involved. There is much misleading information in the literature that suggests that many insect species are very variable in their habits. Such evidence should be treated with caution in the absence of a detailed and competent examination of the populations to demonstrate that they are indeed conspecific and not complexes.

Indeed, host relationships can provide evidence for the existence of unsuspected cryptic species among the host plants themselves. For example, programs aimed at controlling *Lantana camara* have been complicated by the discovery that the species taxon comprises a complex of several forms, some diploid and some tetraploid (Smith and Smith, 1982). The stimulus that led to the revelation of this complex was the study of lantana as a pest and a target for biological control.

During the past twenty years, genetical and cytogenetical tools have come into use, but their utility is much dependent on evolutionary theory. They can be very helpful when used by a specialist evolutionist, but they can prove to be useless or even a handicap in the hands of the inexperienced. The interpretation of results from an electrophoretic study or a cytogenetic study is not easy or obvious. Useful answers are obtained only when critical questions are posed at the planning stage. Generally, these tools will not reveal species automatically; workers planning to use them should understand that the technical ability to run gel electrophoresis or to undertake chromosome studies is only the beginning: the difficult part is the interpretation. The questions to which one seeks answers arise from the genetic theory of species: different genetical concepts dictate different questions, although some may be common. For example, both the isolation concept of Mayr and Dobzhansky and the recognition concept accept that a species is a field for gene recombination. Thus we might ask if the species taxon *H. armigera* at Narrabri constitutes one or more fields of gene recombination. A start might be made by sampling both sunflower and cotton, taking care to avoid biases. If each individual sampled is scored then for, say, fifteen enzyme systems using polyacrylamide gel electrophoresis, we will be in a position to answer the following question: Are these data consistent with the view that they were drawn from a single randomly mating population? If they are statistically homogeneous, then the answer would be yes, and it could be concluded that no reason had been found to doubt that only a single field for gene recombination had been sampled. However, a decision could not be made if no polymorphic characters had been included. If all the individuals from cotton were homozygous for an allele at a particular enzyme locus, and those from sunflower were all homozygous for an alternative allele, one would be forced to conclude that two distinct fields of gene recombination were involved. It might be thought that if the frequencies of alleles at a particular locus are in Hardy-Weinberg equilibrium, then we would have evidence that the sample examined was drawn from a single panmictic (randomly mating) population. However, this is not necessarily so. Suppose two [p. 7] undetected cryptic species, in fact, are present and that both the alleles at this locus are in Hardy-Weinberg equilibrium. A random sample from this mixture of two populations will also be found to be in Hardy-Weinberg equilibrium. Thus, care is needed when using such evidence for inferring that one is dealing with only a single genetical species.

Chromosome cytology can be useful in the detection of cryptic species. In many families of Diptera of economic or medical importance, polytene chromosome studies can be of great value. For example, such chromosomes from the larval salivary glands are used in detecting cryptic species in Simuliidae and *Anopheles*. Polytene chromosomes from the ovarian nurse cells of adult females in some species of the genus *Anopheles* and those from trichogen bristles are valuable in the genera *Calliphora* and *Lucilia*. In other orders of Insecta, one must rely on mitotic and meiotic chromosomes that provide much less detail, even when banding techniques are used. Nevertheless, the chromosomes can still be used to track gene flow as was done by Moran and Shaw (1977) in studying a parapatric zone in the grasshopper genus *Caledia* in Queensland. Karyotypes of related species are often indistinguishable, which means that the chromosomes are sometimes useless as markers to study gene flow. Sometimes it is possible to use morphological markers to study gene flow between populations. However, care is needed to study gene exchange using structural markers, and laboratory crosses and backcrosses should precede their use in the field (Paterson, 1956). Without such preliminary studies, the finding of supposed hybrids in the field generally should be treated skeptically. Frequently, aberrant individuals are considered confidently to be hybrids on the basis of no evidence. Morphological markers are not very useful in studying cryptic species because, generally and by definition, they look very much alike.

DISCUSSION

Bearing these points in mind, it can be understood readily that hitherto underappreciated problems face entomologists, who are concerned with the biological control of weeds and the pests of major crops, as well as insect ecologists specializing in host/predator and host/parasite relationships. Sibling species among such important parasitoid genera as *Trichogramma* are scarcely touched on today. In biological control programs, great care is taken to test prospective controlling species for host range, host specificity, and so on. Yet, how much care is taken to ensure that a complex of species is not treated as a single species or that one sibling species is studied while it is believed to be another? Very few geneticists or entomologists are trained adequately to detect cryptic species. This is a specialized field with few specialists in it.

To illustrate the problems that can result from not detecting cryptic species [p. 8] in an exotic parasitoid on which a control program is to be based, the case of the braconid *Chelonus texanus* can be cited. This was imported into South Africa and released in large numbers in an attempt to control the Karoo caterpillar, *Loxostege frustalis*. The following quote is from the account by Annecke and Moran (1977:139):

> The first release of *C. texanus* commenced in October 1942 and the last consignment was despatched to the Karoo from Pretoria, according to Bedford (1956), on 1 May 1952. During this decade of production of *C. texanus*, more than 8.5 million parasitoids were released . . . against five host species in various parts of the country. In the final year of production, 1,698,000 parasitoids were transferred in the mass-rearing room directly into shipping containers by means of an ingenious light trap designed by Bedford. . . . The parasitoids were sent for release to a total of 491 sites in the Karoo and Orange Free State, and in the two or three years preceding 1950, collections of Karoo caterpillars were made . . . in areas where *C. texanus* had been released. These collections yielded about 2% parasitism by the indigenous *C. curvimaculatus*, a species not then known to attack the Karoo caterpillar, and this species was mistaken for *C. texanus*, which it resembles. According to Bedford (1951), the first of these *Chelonus* was collected in 1943 and reported on (Ullyett, 1944) as *C. texanus*; this error in identification was not corrected until February 1951. The recoveries of the misidentified *C. texanus* were a source of encouragement for the entomologists concerned: Tardrew . . . wrote "This (2% parasitism) may not seem very good, but it is encouraging, showing that the *Chelonus* (*texanus*) can survive the dry conditions of the Karoo, and that it is becoming adapted to its host." . . . Having satisfied himself of the longstanding misidentification of the indigenous *C. curvimaculatus*, Bedford recommended in March 1951 that the mass production programme of *C. texanus* be discontinued at the end of April. His recommendation was overruled, probably for reasons of political expediency, and he was requested to proceed with production and liberation for a further year. . . . In fact none of the parasitoids ever became established and finally, in May 1952, the last liberations of *C. texanus* were made, and the programme was terminated.

It may be thought that this is merely a dreadful example of the way things were once done, but I am not aware of any similar program today that routinely relies on the criteria of genetical species for identification. I believe all still depend on the findings of morphological taxonomy, despite its proven limitations. The apparent discovery of the Oriental fruitfly (*Dacus dorsalis*) in Australia is a recent example of the sort of problems cryptic species can cause (Drew and Hardy, 1981).

Finally, I should like to conclude with the observation that I can think of few advances in insect population biology, basic or applied, that will lead to a greater increase in efficiency than a more careful attendance to the problems arising from cryptic species. By greater efficiency, I do not mean merely the saving of large sums of money across the world, but I mean also [p. 9] that the time of highly trained scientists will be wasted less frequently, the literature will be cluttered up less often with uninterpretable or gravely misleading results, and programs against pests will succeed more often. Detecting sibling species is not just an academic luxury; it is often the difference between success and failure, effectiveness and bumbling.

CONCLUSION

It may be appropriate to make some comments on gaps in our knowledge of *Heliothis* in particular. In order to make critical studies on these organisms it is necessary that we be certain we are studying a single genetic species and that we know exactly which species this is. In the case of *H. armigera* I do not believe we can say that we have done much to ensure that we have met these strictures.

First, a neotype should be designated if the type of Hubner's *Noctua armigera* is, indeed, lost. Evidently, the type was from Europe. Designating a neotype will specify a particular population for reference purposes. Care should be taken then to characterize members of this population structurally, electrophoretically, and biologically. Once this is done, and not before, it will be possible to recognize the typical form wherever it occurs across the world. Similar studies need to be done on the populations at the type localities of *H. armigera conferta* (Auckland, New Zealand) and, less urgently, on that of *Helicoverpa armigera commoni* (Canton Island, South Pacific), which are treated as subspecies of *H. armigera* by Marsh (1978). This is necessary groundwork to provide a firm base for the naming of populations across the world. As pointed out above, taxonomic decisions are not the crucial ones, and neither are electrophoretic ones except in special circumstances. Ultimately, it is the fertilization system under natural conditions that is crucial. With allopatric populations, the best that can be done is to test for assortative mating under conditions that make available simultaneously both forms under test in large cages.

These stringent requirements may be treated with impatience by some. However, if they are not met insecurity will persist. In Australia at present, we do not know really whether what we call *H. armigera* constitutes a single genetic species or a complex. Furthermore, we cannot be quite sure that the true *H. armigera* of Europe is present in Australia. It is also not certain that literature purporting to refer to *H. armigera* in other countries actually refers to any population in Australia. To some this may sound an extreme statement not to be taken seriously. However, I believe that in critical terms it is an accurate reflection of the true situation. At present, we rely on taxonomic decisions that we need to treat with caution, as has been demonstrated with the analysis of such other taxa as *Anopheles gambiae*. [p. 10]

The situation with *H. punctigera* is somewhat less difficult, because it is an endemic species. However, it too should be studied in detail at its type locality (Sydney) to provide a sound basis for detailed studies across the continent.

ACKNOWLEDGMENTS

I am pleased to acknowledge assistance I have received from Drs. Rachel McFadyen, Bill Palmer, Gimme Walter, David Yates, and Myron Zalucki, and, particularly, Ian Common.

REFERENCES

Annecke, D. P., and V. C. Moran. 1977. Critical reviews of biological pest control in South Africa, I. The Karoo caterpillar, *Loxostege frustalis* Zeller (Lepidoptera: Pyralidae). *Journal of the Entomological Society of Southern Africa* 40:127–45.

Ayala, F. J., M. L. Tracey, D. Hedgecock, and R. C. Richmond. 1974. Genetic differentiation during the speciation process in *Drosophila*. *Evolution* 28:576–92.

Bedford, E.C.G. 1951. Unpublished report. Pretoria: Plant Protection Research Institute.

———. 1956. The automatic collection of mass-reared parasites into consignment boxes, using two light sources. *Journal of the Entomological Society of Southern Africa* 19:342–53.

Cardé, R. T., W. L. Roelofs, R. G. Harrison, A. T. Vawter, P. F. Brussard, A. Mutuura, and E. Munroe. 1978. European corn borer: Pheromone polymorphism or sibling species? *Science* 199:555–56.

Carson, H. L. 1957. The Species as a Field for Gene Recombination. In *The Species Problem*, ed. E. Mayr, 23–38. Publication No. 50. Washington, D.C.: American Association for the Advancement of Science.

Daly, J. C. 1991. Methods for Studying the Genetics of Populations of *Heliothis*.

In Heliothis: *Research Methods and Prospects*, ed. M. P. Zalucki, 157–70. New York: Springer-Verlag.

Daly, J. C., and P. Gregg. 1985. Genetic variation in *Heliothis* in Australia: Species identifications and gene flow in the two pest species *H. armigera* (Hübner) and *H. punctigera* Wallengren (Lepidoptera: Noctuidae). *Bulletin of Entomological Research* 75:169–84.

Davidson, G. 1956. Insecticide resistance in *Anopheles gambiae* Giles: A case of simple Mendelian inheritance. *Nature* 178:863–64.

———. 1958. Studies on insecticide resistance—resistance in *Anopheles* mosquitoes. *Bulletin of the World Health Organisation* 16:579–621.

Dobzhansky, T. 1951. *Genetics and the Origin of Species*. 3d ed. New York: Columbia University Press.

Drew, R.A.I., and D. E. Hardy. 1981. *Dacus (Bactrocera) opiliae*, a new sibling species of the *dorsalis* complex of fruit flies from northern Australia (Diptera: Tephritidae). *Journal of the Australian Entomological Society* 20:131–37.

Holstein, M. H. 1957. Cytogenetics of *Anopheles gambiae*. *Bulletin of the World Health Organisation* 16:456–58.

Johnson, W. E., H. L. Carson, K. Y. Kaneshiro, W.W.M. Steiner, and M. M. Cooper. 1975. Genetic Variation in Hawaiian *Drosophila*, II. Allozymic Differentiation in the *D. planitibia* Subgroup. In Isozymes. *Genetics and Evolution*, ed. C. L. Markert, 563–84. Vol. 4. New York: Academic Press.

Key, K.H.L. 1976. A generic and suprageneric classification of the Morabinae (Orthoptera: Eumastacidae), with descriptions of the type species and a bibliography of the subfamily. *Australian Journal of Zoology, Supplementary Series* 37:1–185.

Lambert, D. M., and H.E.H. Paterson. 1982. Morphological resemblance and its relationship to genetic distance measures. *Evolutionary Theory* 5:291–300. [This volume, chap. 7.]

Mahon, R. J., P. M. Miethke, and J. A. Mahon. 1982. The evolutionary relationships of the jarrah leaf miner, *Perthida glyphopa* (Common) (Lepidoptera: Incurvariidae). *Australian Journal of Zoology* 30:243–49.

Marsh, P. M. 1978. The braconid parasites (Hymenoptera) of *Heliothis* species (Lepidoptera: Noctuidae). *Proceedings of the Entomological Society of Washington* 80:15–36.

Mayr, E. 1942. *Systematics and the Origin of Species*. New York: Columbia University Press.

———. 1963. *Animal Species and Evolution*. Cambridge, Mass.: Belknap Press of Harvard University Press.

Moran, C., and D. D. Shaw. 1977. Population cytogenetics of the genus *Caledia* (Orthoptera: Acridinae), III. Chromosomal polymorphism, racial parapatry and introgression. *Chromosoma (Berlin)* 63:181–204.

Muirhead-Thomson, R. C. 1948. Studies on *Anopheles gambiae* and *A. melas* in and around Lagos. *Bulletin of Entomological Research* 38:527–58.

———. 1951. Studies on salt-water and fresh-water *Anopheles gambiae* on the East African coast. *Bulletin of Entomological Research* 41:487–502.

Paterson, H.E.H. 1956. Status of the two forms of housefly occurring in South Africa. *Nature* 178:928–29.

———. 1962. Status of the East African salt-water-breeding variant of *Anopheles gambiae* Giles. *Nature* 195:469–70.

———. 1978. More evidence against speciation by reinforcement. *South African Journal of Science* 74:369–71. [This volume, chap. 3.]

———. 1982. Perspective on speciation by reinforcement. *South African Journal of Science* 78:53–57. [This volume, chap. 8.]

———. 1985. The Recognition Concept of Species. In *Species and Speciation*, ed. E. S. Vrba, 21–29. Transvaal Museum Monograph No. 4. Pretoria: Transvaal Museum. [This volume, chap. 12.]

Smith, L. S., and D. A. Smith. 1982. The naturalized *Lantana camara* complex in eastern Australia. *Qld. Bot. Bull.* 1:1–26.

Sokal, R. R. 1973. The species problem reconsidered. *Systematic Zoology* 22:360–74.

Tinbergen, N., B.J.D. Meeuse, L. K. Boerema, and W. W. Varossieau. 1942. Die Balz des Samfalters, *Eumenis (Satyrus) semele* (L.). *Zeitschrift für Tierpsychologie* 5:182–226.

Ullyett, G. C. 1944. Unpublished annual report of the Parasite Laboratory, Pretoria: Plant Protection Research Institute.

Whittle, C. P., T. E. Bellas, M. Horak, and B. Pinese. 1987. The sex pheromone and taxonomic status of *Homona spargotis* Meyrick sp. rev., an Australian pest species of the *Coffearia* group (Lepidoptera: Tortricidae). *Journal of the Australian Entomological Society* 26:169–79.

Complete Bibliography of
H.E.H. Paterson

Boyes, J. W., M. J. Corey, and H.E.H. Paterson. 1964. Somatic chromosomes of higher Diptera, IX. Karyotypes of some Muscid species. *Canadian Journal of Zoology* 42:1025–36.

Davidson, G., H.E.H. Paterson, M. Coluzzi, G. F. Mason, and D. W. Micks. 1967. The *Anopheles gambiae* Complex. Chap. 6 in *Genetics of Vectors of Disease*, ed. J. W. Wright and R. Pal, 211–250. Amsterdam: Elsevier.

de Meillon, B., H.E.H. Paterson, and J. Muspratt. 1957. Notes on the more common mosquitoes of Tongaland. *South African Journal of Medical Science* 22:41–46.

Irving-Bell, R., and H.E.H. Paterson. 1973. Intracellular symbionts and infertility in mosquitoes of the *Culex pipiens* complex. Never published although cited by Paterson and James, 1973, as in press.

Kokernot, R. H., B. de Meillon, H.E.H. Paterson, C. S. Heymann, and K. S. Smithburn. 1957. Middelburg virus, a hitherto unknown agent isolated from *Aedes* mosquitoes during an epizootic in sheep in the Eastern Province. *South African Journal of Medical Science* 22:145–53.

Kokernot, R. H., H.E.H. Paterson, and B. de Meillon. 1958. Studies on the transmission of Wesselsbron virus by *Aedes* (*Ochlerotatus*) *caballus* (Theo.). *South African Medical Journal* 32:546–48.

Kokernot, R. H., K. C. Smithburn, H.E.H. Paterson, and B. de Meillon. 1960. Further isolations of Wesselsbron virus from mosquitoes. *South African Medical Journal* 34:871–74.

———. 1960. Isolation of Germiston virus, a hitherto unknown agent, from Culicine mosquitoes, and a report of infection in two laboratory workers. *American Journal of Tropical Medicine and Hygiene* 9:62–69.

Kokernot, R. H., K. D. Smithburn, B. de Meillon, and H.E.H. Paterson. 1958. Isolation of Bunyamwera virus from a naturally infected human being, and further isolations from *Aedes* (*Banksinella*) *circumluteolus* Theo. *American Journal of Tropical Medicine and Hygiene* 7:579–84.

Lambert, D. M., M. R. Centner, and H.E.H. Paterson. 1984. Simulation of the conditions necessary for the evolution of species by reinforcement. *South African Journal of Science* 80:308–11. [This volume, chap. 11.]

Lambert, D. M., and H.E.H. Paterson. 1982. Morphological resemblance and its relationship to genetic distance measures. *Evolutionary Theory* 5:291–300. [This volume, chap. 7.]

———. 1984. On "Bridging the gap between race and species": The isolation concept and an alternative. *Proceedings of the Linnean Society of New South Wales* 107:501–14.

Masters, J., D. M. Lambert, and H.E.H. Paterson. 1984. Scientific prejudice, reproductive isolation, and apartheid. *Perspectives in Biology and Medicine* 28:107–16.

McIntosh, B. M., R. M. Harwin, H.E.H. Paterson, and M. L. Westwater. 1963. An epidemic of Chikungunya in South-eastern Southern Rhodesia. *Centr. Afr. Med. J.* 9:351–59.

McIntosh, B. M., H.E.H. Paterson, J. M. Donaldson, and J. de Sousa. 1963. Chikungunya virus: Viral susceptibility and transmission studies with some vertebrates and mosquitoes. *South African Journal of Medical Science* 28:45–52.

McIntosh, B. M., H.E.H. Paterson, G. McGillivray, and J. de Sousa. 1963. Further studies on the Chikungunya outbreak in Southern Rhodesia in 1962, I. Mosquitoes, wild primates and birds in relation to the epidemic. *Ann. Trop. Med. Parasit.* 58:45–51.

Miles, S. J., and H.E.H. Paterson. 1979. Protein variation and systematics in the *Culex pipiens* group of species. *Mosquito Systematics* 11:187–202.

Muspratt, J., K. C. Smithburn, H.E.H. Paterson, and R. H. Kokernot. 1957. The laboratory transmission of Wesselsbron virus by the bite of *Aedes* (*Banksinella*) *circumluteolus* Theo. *South African Journal of Medical Science* 22:121–26.

Paterson, H.E.H. 1953. Bird-lice and evolution. *Bokmakierie* 5:14–15.

———. 1953. *Fannia albitarsis* Stein—a species new to the Ethiopian region. *Journal of the Entomological Society of Southern Africa* 16:79.

———. 1953. New *Lispe* species (Diptera, Muscidae) from Southern Africa. *Journal of the Entomological Society of Southern Africa* 16:168–78.

———. 1953. A new species of the genus *Puliciphora* Dahl (Diptera: Phoridae). *Revista Ecuatoriana de Entomologia y Parasitologia* 1:61–67.

———. 1954. A new record of *Pedicinus hamadryas* Mjoberg (Phthiraptera, Anoplura) from the Chacma Baboon in South Africa. *Journal of the Entomological Society of Southern Africa* 17:139.

———. 1954. A new record of the quill-boring habit in Mallophaga. *The Entomologist's Monthly Magazine* 90:158.

———. 1954. New sucking lice from South Africa. *Revista Ecuatoriana de Entomologia y Parasitologia* 2:219–25.

———. 1955. Two new species of *Limnophora* (Diptera, Muscidae) related to *Limnophora transludica* Stein. *Journal of the Entomological Society of Southern Africa* 18:137–43.

———. 1956. East African Muscidae (Diptera). *Beiträge zur Entomologie, Berlin* 6:154–79.

———. 1956. Status of the two forms of housefly occurring in South Africa. *Nature* 178:928–29.

———. 1957. A new genus and two new species of Muscini from South Africa (Diptera, Muscidae). *Journal of the Entomological Society of Southern Africa* 20:445–49.

———. 1957. Note on wildfowl parasites in South Africa. *Bokmakierie* 9:48–49.

———. 1957. Three new species of *Musca* from Southern Africa. *Journal of the Entomological Society of Southern Africa* 20:106–13.

———. 1958. A new *Pyrellia* species from Natal, together with miscellaneous notes on other Muscidae. *Journal of the Entomological Society of Southern Africa* 21:300–305.

———. 1958. Sex linked and sex limited mutation of the fly *Muscina stabulans* (Fal.). *Nature* 181:932–33.

———. 1958. *Stygeromyia zumpti*, a new bloodsucking fly from the Transvaal (Diptera, Muscidae). *Journal of the Entomological Society of Southern Africa* 21:80–84.

———. 1959. Notes on the genus *Alluaudinella* G.-T. with the description of a new species and a key to the known species of the genus (Diptera, Muscidae). *Mem. Inst. Sci. Madagas.* 11:355–67.

———. 1960. Diptera (Brachycera, Muscidae): Muscinae and Lispinae. In *Animal Life in South Africa*, ed. P. Brinck, 397–401. Vol. 7. Stockholm: Ålmqvist and Wiksell.

———. 1960. Sibling species in the *melanogaster* species-group in Africa. *Drosophila Information Service* 34:100–101.

———. 1961. Annual Report. East African Institute of Malaria and Vector-borne Diseases, 1960–1961.

———. 1962. On the status of the East African saltwater-breeding variant of *Anopheles gambiae* Giles. WHO/Mal./346. Unpublished document.

———. 1962. Status of the East African salt-water-breeding variant of *Anopheles gambiae* Giles. *Nature* 195:469–70.

———. 1963. On the naming of the indigenous houseflies of the Ethiopian region. *Journal of the Entomological Society of Southern Africa* 26:226–27.

———. 1963. The practical importance of recent work on the *Anopheles gambiae* complex. Seventh International Congress of Tropical Medicine and Malaria, Rio de Janeiro.

———. 1963. Recent investigations of the *Anopheles gambiae* complex in Southern Africa, I. & II. On the naming of the East African saltwater species of the *Anopheles gambiae* complex. *WHO Document on Malaria* 421.

———. 1963. The species, species control and antimalarial spraying campaigns: Implications of recent work on the *Anopheles gambiae* complex. *South African Journal of Medical Science* 28:33–44.

———. 1964. Direct evidence for the specific distinctness of forms A, B, and C of the *Anopheles gambiae* complex. *Rivista di Malariologia* 43:191–96.

———. 1964. Population Genetic Studies in Areas of Overlap of Two Subspecies of *Musca domestica* L. In *Ecological Studies in Southern Africa*, ed. D.H.S. Davis, 244–54. The Hague: W. Junk.

———. 1964. "Saltwater *Anopheles gambiae*" on Mauritius. *Bulletin of the World Health Organisation* 31:635–44.

———. 1965. Direct evidence for the specific distinctness of forms A, B and C of the *Anopheles gambiae* complex. *Entomologie médicale* 3–4:179–80.

———. 1968. Evolutionary and Population Genetical Studies of Certain Diptera. Ph.D. diss., University of the Witwatersrand, Johannesburg.

————. 1973. Animal species studies. Pp. 31–36 in Animal and plant speciation studies in Western Australia, by H.E.H. Paterson and S. H. James. *Journal of the Royal Society of Western Australia* 56:31–43. [See this volume, Author's Preface.]

————. 1974. Mosquitoes. In *Dam-induced Ecological Disturbances Affecting Human Health*, ed. N. R. Stanley and M. P. Alpers, 301–9. London: Academic Press.

————. 1975. The *Musca domestica* complex in Sri Lanka. *Journal of Entomology (B)* 43:247–59. [Dated 1974, printed 7 January 1975.]

————. 1976. The term "Isolating Mechanisms" as a canalizer of evolutionary thought. Unpublished MS. [This volume, chap. 1.]

————. 1976. The role of postmating isolation in evolution. Invited Lecture Fifteenth International Congress of Entomology, Washington, Symposium on the Application of Genetics to the Analyses of Species Differences. Unpublished MS. [This volume, chap. 2.]

————. 1978. Education through Animals: Inaugural Lecture. Wiwatersrand University Press, Johannesberg. 11 pages.

————. 1978. More evidence against speciation by reinforcement. *South African Journal of Science* 74:369–71. [This volume, chap. 3.]

————. 1980. A comment on "Mate Recognition Systems." *Evolution* 34:330–31. [This volume, chap. 4.]

————. 1981. The continuing search for the unknown and unknowable: A critique of contemporary ideas on speciation. *South African Journal of Science* 77:113–19. [This volume, chap. 5.]

————. 1981. Epitaph to a scientific gadfly. *South African Journal of Science* 77:195–96. [This volume, chap. 6.]

————. 1982. Perspective on speciation by reinforcement. *South African Journal of Science* 78:53–57. [This volume, chap. 8.]

————. 1982. Darwin and the origin of species. *South African Journal of Science* 78:272–75. [This volume, chap. 9.]

————. 1985. The Recognition Concept of Species. In *Species and Speciation*, ed. E. S. Vrba, 21–29. Transvaal Museum Monograph No. 4. Pretoria: Transvaal Museum. [This volume, chap. 12.]

————. 1986. Environment and species. *South African Journal of Science* 82:62–65. [This volume, chap. 13.]

————. 1987. A view of species. *Rivista di Biologia–Biology Forum* 80:211–15.

————. 1988. On defining species in terms of sterility: Problems and alternatives. *Pacific Science* 42:65–71. [This volume, chap. 14.]

————. 1989. A View of Species. Chap. 6 in *Dynamic Structures in Biology*, ed. B. Goodwin, A. Sibatani, and G. Webster, 77–88. Edinburgh: Edinburgh University Press. [This volume, chap. 15.]

————. 1989. Updating the evolutionary synthesis. *Rivista di Biologia–Biology Forum* 82:371–75. [This volume, chap. 16.]

————. 1991. The Recognition of Cryptic Species among Economically Important Insects. Chap. 1 in Heliothis: *Research Methods and Prospects*, ed. M. P. Zalucki, 1–10. New York: Springer-Verlag. [This volume, chap. 17.]

Paterson, H.E.H., P. Bronsden, J. Levitt, and C. B. Worth. 1964. Some culicine mosquitoes (Diptera, Culicidae) at Ndumu, Republic of South Africa: A study of their host preferences and host range. Medical Proceedings 10:188–92.

Paterson, H.E.H., C. A. Green, and R. J. Mahon. 1976. *Aedes aegypti* complex in Africa. *Nature* 259:252.

Paterson, H.E.H., and S. H. James. 1973. Animal and plant speciation studies in Western Australia. *Journal of the Royal Society of Western Australia* 56:31–43. [See this volume, Author's Preface.]

Paterson, H.E.H., R. H. Kokernot, and D.H.S. Davis. 1957. The birds of Tongaland and their possible role in virus disease. *South African Journal of Medical Science* 22:63–69.

Paterson, H.E.H., and M. Macnamara. 1984. The recognition concept of species (Michael Macnamara interviews H.E.H. Paterson). *South African Journal of Science* 80:312–18. [This volume, chap. 10.]

Paterson, H.E.H., and B. M. McIntosh. 1963. Further studies on the Chikungunya outbreak in Southern Rhodesia in 1962, II. Transmission experiments with the *Aedes furciferitaylori* group of mosquitoes, and with a member of the *Anopheles gambiae* complex. *Ann. Trop. Med. Parasit.* 58:52–55.

Paterson, H.E.H., and K. R. Norris. 1970. The *Musca sorbens* complex: The relative status of the Australian and two African populations. *Australian Journal of Zoology* 18:231–45.

Paterson, H.E.H., J. S. Paterson, and G. J. van Eeden. 1963. A new member of the *Anopheles gambiae* complex: A preliminary report. *Medical Proceedings* 9:414–18.

———. 1964. Records of the breeding of saltwater *Anopheles gambiae* at inland localities in southern Africa. *Nature* 201:524–25.

Paterson, H.E.H., and P. M. Thompson. 1953. A key to the Ethiopian species of the genus *Polyplax* (Anoplura), with descriptions of two new species. *Parasitology* 43:199–204.

Paterson, H.E.H., and L. Tsacas. 1967. The identification of *Drosophila seguyi* Smart. *Drosophila Information Service* 42:73.

Paterson, H.E.H., and C. B. Worth. 1961. Gynandromorphism in an African mosquito. *Journal of the Entomological Society of Southern Africa* 24:214–15.

Pont, A., and H.E.H. Paterson. 1971. The Genus *Musca*. In *Flies and Disease*. Ecology, Classification and Biotic Associations, ed. B. Greenberg. Vol. 1. 108–115. Princeton, N.J.: Princeton University Press.

Robertson, H. M., and H.E.H. Paterson. 1982. Mate recognition and mechanical isolation in *Enallagma* damselflies (Odonata: Coenagrionidae). *Evolution* 36:243–50.

Smithburn, K. C., H.E.H. Paterson, C. S. Heymann, and P. A. Winter. 1959. An agent related to Uganda S virus from man and mosquitoes in South Africa. *South African Medical Journal* 33:959–62.

Weinbren, M. P., C. Heymann, R. H. Kokernot, and H.E.H. Paterson. 1957. Simbu virus, a hitherto unknown agent isolated from *Aedes* (*Banksinella*) *circumluteolus* Theo. *South African Journal of Medical Science* 22:93–102.

Worth, C. B., and H.E.H. Paterson. 1960. Phoresy of sucking lice by a mosquito. *Journal of the Entomological Society of Southern Africa* 22:228–30.

———. 1961. Culicine mosquitoes in southern Africa. New locality records in the province of Natal (Union of South Africa), Mozambique (Portuguese East Africa) and Angola (Portuguese West Africa), with a list of species now known from Natal. *Rev. Ent. Moçambique* 4:65–80.

Worth, C. B., H.E.H. Paterson, and B. de Meillon. 1961. The incidence of arthropod-borne viruses in a population of Culicine mosquitoes in Tongaland, Union of South Africa (January 1956, through April 1960). *American Journal of Tropical Medicine and Hygiene* 10:583–92.

Zumpt, F., and H.E.H. Paterson. 1953. Studies on the family Gastrophilidae, with keys to the adults and maggots. *Journal of the Entomological Society of Southern Africa* 16:56–72.

Supplemental References

Bastock, M. 1956. A gene mutation which changes a behavior pattern. *Evolution* 10:421–39.

Blair, W. S. 1955. Mating call and stage of speciation in the *Microhyla olivacea–M. cardinensis* complex. *Evolution* 9:469–80.

Butlin, R. K. 1987. Species, speciation, and reinforcement. *American Naturalist* 130:461–64.

Dobzhansky, T. 1950. Mendelian populations and their evolution. *American Naturalist* 84:401–18.

———. 1976. Organismic and Molecular Aspects of Species Formation. In *Molecular Evolution*, ed. F. J. Ayala, 95–105. Sunderland, Mass.: Sinauer Associates.

Ehrman, L. 1972. Genetics and Sexual Selection. In *Sexual Selection and the Descent of Man 1871–1971*, ed. B. Campbell, 105–35. London: Heinemann.

Eldredge, N. 1989. *Macroevolutionary Dynamics: Species, Niches, and Adaptive Peaks*. New York: McGraw-Hill.

Fouquette, M. J. 1975. Speciation in chorus frogs, I. Reproductive character displacements in the *Pseudacris nigrita* complex. *Systematic Zoology* 24:16–22.

Haldane, J.B.S. 1955. Some alternatives to sex. *New Biology* 19:7–26.

Harper, A. A., and D. M. Lambert. 1983. The population genetics of reinforcing selection. *Genetica* 62:15–23.

Henderson, N. R., and D. M. Lambert. 1982. No significant deviation from random mating of worldwide populations of *Drosophila melanogaster*. *Nature* 300:437–40.

Kalmus, H. 1941. The resistance to desiccation of *Drosophila* mutants affecting body colour. *Proceedings of the Royal Society* 130:185–201.

Lambert, D. M., M. R. Centner, and H.E.H. Paterson. 1984. Simulation of the conditions necessary for the evolution of species by reinforcement. *South African Journal of Science* 80:308–11. [This volume, chap. 11.]

Lambert, D. M., and H.E.H. Paterson. 1984. On "Bridging the gap between race and species": The isolation concept and an alternative. *Proceedings of the Linnean Society of New South Wales* 107:501–14.

Levin, D. A. 1970. Reinforcement of reproductive isolation: Plants versus animals. *American Naturalist* 104:571–81.

Littlejohn, M. J. 1981. Reproductive Isolation: A Critical Review. Chap. 15 in *Evolution and Speciation*, ed. W. R. Atchley and D. S. Woodruff, 298–334. Cambridge: Cambridge University Press.

Margulis, L. 1970. *Origin of Eukaryotic Cells*. New Haven, Conn.: Yale University Press.

219

Masters, J., D. M. Lambert, and H.E.H. Paterson. 1984. Scientific prejudice, reproductive isolation, and apartheid. *Perspectives in Biology and Medicine* 28:107–16.

Mayr, E. 1942. *Systematics and the Origin of Species.* New York: Columbia University Press.

———. 1949. Speciation and Systematics. In *Genetics, Paleontology, and Evolution,* ed. G. L. Jepsen, E. Mayr, and G. G. Simpson, 281–98. Princeton, N.J.: Princeton University Press.

———. 1963. *Animal Species and Evolution.* Cambridge, Mass.: Belknap Press of Harvard University Press.

———. 1969. *Principles of Systematic Zoology.* New York: McGraw–Hill.

———. 1978. Origin and history of some terms in systematic and evolutionary biology. *Systematic Zoology* 27:83–88.

Moore, J. A. 1957. An Embryologist's View of the Species Concept. In *The Species Problem,* ed. E. Mayr, 325–38. Publication No. 50. Washington, D.C.: American Association for the Advancement of Science.

Morton, E. S. 1975. Ecological sources of selection on avian sounds. *American Naturalist* 109:17–34.

Noble, G. K. 1936. Courtship and sexual selection of the flicker (*Colaptes auratus leuteus*). *Auk* 53:269–82.

Petersen, W. 1905. Uber beginnende Art-Divergenz. *Arch. Rass. u. Ges. Biol.* 2:641–62.

Ralin, D. B. 1977. Evolutionary aspects of mating call variation in a diploid-tetraploid species complex of tree frogs (Anura). *Evolution* 31:721–37.

Robertson, H. M., and H.E.H. Paterson. 1982. Mate recognition and mechanical isolation in *Enallagma* damselflies (Odonata: Coenagrionidae). *Evolution* 36:243–50.

Spencer, H. G., D. M. Lambert, and B. H. McArdle. 1987. Reinforcement, species, and speciation: A reply to Butlin. *American Naturalist* 130:958–62.

Spencer, H. G., B. H. McArdle, and D. M. Lambert. 1986. A theoretical investigation of speciation by reinforcement. *American Naturalist* 128:241–62.

Templeton, A. R. 1979. Once again, why 300 species of Hawaiian *Drosophila? Evolution* 33:513–17.

Verrell, P. A. 1988. Stabilizing selection, sexual selection and speciation: A view of specific-mate recognition systems. *Systematic Zoology* 37:209–15.

Taxonomic Index

adiastola subgroup (*Drosophila*), 68
Aedes stimulans (mosquito), 186
Algae, 137; brown, 120; green, 101
Allium schoenoprasum (chives), 61
Amoeba, 189
Anas, 26; *platyrhynchos* (mallard), 26, 174; *acuta* (pintail duck), 26, 174
Anatini (ducks), 26, 146
Anopheles (mosquito), xiii, 206; *arabiensis*, 204; *gambiae*, 203–4, 209; *gambiae* complex, xiii, 69–70; *marshallii*, 69–70; *melas*, 203; *merus*, 203
Antelope, 87
Anura, 199
Apes, 71, 199
Apple, 46, 49, 51
Apply fly. See *Rhagoletis arizonensis*, *Drosophila*, 25
Avocado (*Persea*), 204

Bacterium (Rickettsia), 16–17
Bat, 146
Beetle (Coleoptera), 48, 81
biloba. See *Clarkia*
Bird-of-paradise, 183
Blackfly (Simuliidae), 206
Blowfly. See *Calliphora*; *Chrysomya*; *Lucilia*
Braconid (wasp), 207
Bufo (toad): *americanus*, 26; *woodhousii fowleri*, 26

Caledia (grasshopper), 25, 206; *captiva*, 25
Calliphora (blowfly), 206
Cepaea nemoralis (mollusk), 42
Chelonus (wasp), 207; *curvimaculatus*, 207; *texanus*, 207
Cherry fly. See *Rhagoletis*
Chimpanzee, 71
Chrysomya (blowfly), xiii
Clarkia (angiosperm), 25, 44; *biloba*, 44, 168, 173, 176; *lingulata*, 44, 168, 173–74, 176
Colaptes (flicker, woodpecker), 77
Coleoptera, 87
Cotoneaster (Roseaceae), 48
Crataegus (hawthorn, Roseaceae), 46, 48
Cricket (*Gryllus*, *Teleogryllus*), 27

Crustacea, 199
Cryptostylus ovata (slipper orchid), 147
Culex (mosquito), xiii, 17, 65; *australicus*, 17; *globocoxitus*, 17; *molestus*, 16–17, 51, 65, 150; *pallens*, 17; *pipiens*, 17, 150; *pipiens* complex, 16–18, 176; *quinquefasciatus*, 16–17

Dacus dorsalis (fruitfly, Tephritidae), 45, 208
Dodo (*Raphus*), 35
Drosophila, 22, 32–33, 63, 68, 70, 84, 113–14; *adiastola* subgroup, 68; *arizonensis*, 25; *cyrtoloma*, 68; *equinoxialis caribbensis*, 23; *equinoxialis equinoxialis*, 23; *heteroneura*, 68, 202; *ingens*, 68; *melanocephala*, 68; *melanogaster*, xvi, 26, 66–67, 84, 128, 130–31, 163; *mojavensis*, 25; *ochrobasis*, 68; *paulistorum*, 51, 150, 176; *paulistorum* complex, 18; *persimilis*, 25, 84; *planitibia* subgroup, 68; *pseudoobscura*, 25, 66, 84; *setosimentum*, 68; *silvestris*, 68, 202; *willistoni quechua*, 23; *willistoni willistoni*, 23
Duck. See *Anas*; Anatini

Eucalyptus: *marginata* (jarrah), 203; *rudis* (flooded gum), 203; *todtiana* (prickly bark gum), 203
Euglenoids, 189

Finch, Cocos Island (*Pinaroloxias*), 8–9
Fish. See *Gasterosteus*; *Pygosteus*
Flicker (woodpecker, *Colaptes*), 77
Frog. See *Heleophryne*; *Hyla*; *Litoria*; *Xenopus*
Fruitfly, oriental (*Dacus*, Tephritidae), 45, 208

gambiae. See *Anopheles*
Gasterosteus acuta (three-spined stickleback fish), 118, 146. See also *Pygosteus*
Geocolaptes olivaceus (ground woodpecker), 164–65
Grasshopper. See *Caledia*; *Warramaba*
Gryllus (cricket), 27

Gum (*Eucalyptus*): flooded, 203; prickly
   bark, 203

Hamerkop (*Scopus*), 9, 154
Hammerhead, African (*Scopus*), 9, 154
Hawthorn (*Crataegus*), 46, 48
Hawthorn fly. See *Rhagoletis*
Heleophryne purcelli (Cape Ghost frog),
   86
Helicoverpa, subgenus. See *Heliothis*
Heliothis (noctuid moth), 200, 208;
   armigera, 204–5, 208–9, *armigera
   commoni*, 208; *armigera conferta*, 208;
   coffearia, 204; *punctigera*, 209
Homona spargotis (tortricid moth), 204
Homo sapiens, 71, 93, 116, 153–54, 181–
   82, 202
Humans. See *Homo*
Hummingbird, ruby-throated, 184
Hyla (frog), 71; *eximia*, 70; *regilla*, 70.
   See also *Litoria*

Jarrah (*Eucalyptus*), 203

Karoo caterpillar (*Loxostege*), 207

Lantana camara, 205
Lepidoptera, 199
lingulata. See *Clarkia*
Litoria (frog): *ewingi*, 82; *verreauxi*, 82.
   See also *Hyla*
Loxostege frustalis (Karoo caterpillar), 207
Lucilia (blowfly), 206

Madagascan partridge (*Margaroperdix*),
   8, 154
Mallard (*Anas*), 26, 174
Malus (apple), 46, 49, 51
Man. See *Homo*
Maniola jurtina (meadow brown
   butterfly), 42
Margaroperdix madagascariensis
   (Madagascan partridge), 9, 154
Mastomys natalensis (multimammate
   mouse), 71
Metazoa, 6, 13
Microcryptorhynchus (weevil), 48
Minnow, Californian, 71
Mollusk. See *Cepaea*; *Nucella*
Morabine grasshopper (*Warramaba*), 53,
   201
Mosquito. See *Aedes*; *Anopheles*; *Culex*
Moth. See *Heliothis*; *Homona*; *Noctua*;
   Ostrinia; *Perthida*; *Solenobia*
Musca (fly), xiii
Myxotricha paradoxa (protistan), 195

Noctua armigera (moth), 208
Nucella lapillus (dog whelk, mollusk), 45

Orchid (*Cryptostylus*), 147
Ostrich (*Struthio*), 154, 184
Ostrinia nubilalis (European corn-borer
   moth), 8, 69
Oyster, 120

Parus caeruleus complex (tit), 26
Pavo cristatus (peacock, peafowl), 100,
   182–83
Peacock (*Pavo*), 182–83
Peafowl (*Pavo*), 100
Pelecaniformes (Aves), 184
Persea americana (avocado), 204
Perthida glyphopa (jarrah leaf-miner
   moth), 203
Pezophaps (solitaire, Aves), 35
Pinaroloxias inornata (Cocos Island finch),
   8–9
Pintail duck (*Anas*), 26, 174
Plague bacilli, 71
planitibia, subgroup (*Drosophila*), 68
Prickly bark gum (*Eucalyptus*), 203
Primates, 71, 199
Protista, 196
Protozoa, 100
Prunus (Roseaceae), 48
Pygosteus pungitius (ten-spined stickleback
   fish), 146. See also *Gasterosteus*
Pyrus (Roseaceae), 48

Raphus cucullatus (dodo), 35
Rats, mole, 71
Rhagoletis pomonella (apple or cherry fly),
   45–46, 48
Rickettsia (bacterium), 16–17; as
   symbiont (see *Wohlbachia*)
Rodentia, 199

Sarcophaga haemorrhoidalis (fleshfly), 118
Scopus umbretta (African hammerhead or
   hamerkop), 9, 154
Sepia officinalis (squid), 146
Simuliidae (blackfly), 206
Snake, ribbon, 71
Solenobia (psychid moth), 51
Solitaire (*Pezophaps*), 35
Squid (*Sepia*), 146
Stickleback fish. See *Gasterosteus*;
   Pygosteus
Struthio camelus (ostrich), 154, 184

Teleogryllus (cricket), 27
Ten-spined stickleback. See *Pygosteus*
Tephritidae (fruitfly, *Dacus*), 45

Three-spined stickleback. See
  Gasterosteus
Tit (*Parus*), 26
Toad. See *Bufo*
Tree frog, North American. See *Hyla*
Trichogramma (wasp), 206

Vandiemenella: *viatica*, 43; *viatica*
  complex, 42–44
verreauxi, *Litoria*, 82

Warramaba virgo (morabine
  grasshopper), 53, 201
Wasp. *See* Braconid; *Chelonus*;
  Trichogramma
Weevil (*Microcryptorhynchus*), 48
Wohlbachia pipientis (Rickettsia), 16, 51
Woodpecker. See *Colaptes*; *Geocolaptes*

Xenopus laevis (African clawed frog), 77

Zea mays, 25, 84

Author Index

Titles of all published works (journal articles, articles in edited volumes, and books) are in italics.
Page numbers refer to pages on which author's name or author/date citation appears.

Alexander, R. D.: cited by Loftus-Hills, 78
Anderson, E., 184
Anderson, J. F., 186
Anderson, W. W., 66
Andrewartha, H. G., 151
Annecke, D. P., 207
Antonovics, J., 126
Avise, J. C., 64, 68, 71
Ayala, F. J.: *Adaptive differentiation with little genic change between two native Californian minnows* (1975), 71; *Genetic differentiation during the speciation process* (1975), 64, 67–68, 125; *Genetic differentiation during the speciation process in* Drosophila (1977), 22–23, 25–27, 78, 83, 202; *Humankind: A Product of Evolutionary Transcendence* (1977), 195; *Humans and apes are genetically very similar* (1978), 71

Barr, A. R., 16
Bastock, M., xvi, 27, 67
Bazykin, A. D., 23
Bedford, E.C.G.: cited by Annecke and Moran, 207
Bellas, T. E., 204
Bentley, D. R., 27
Birch, L. C., 151
Blackith, R. E., 42
Blackith, R. M., 42
Blair, W. S., 79; cited by Littlejohn, 78
Blyth, E.: quoted from Eiseley, 95
Bocquet, C., 66
Boerema, L. K., 203
Bossert, W. H., 27, 126
Braithwaite, R. B., 108
Brown, K., 71
Bruce, E. J., 71
Brussard, P. F., 69–70, 204
Bush, G. L., 22, 45–48, 54, 81, 159
Butlin, R. K., 168

Cain, A. J., 1
Caisse, M., 126
Cardé, R. T., 69–70, 204

Carson, H. L.: *Allozymic and chromosomal similarity in two* Drosophila *species* (1975), 68, 202; *Chromosomes and species formation* (1974), 42, 48; *The Evolutionary Biology of the Hawaiian Drosophilidae* (1970), 68; *The genetics of speciation at the diploid level* (1975), 70; *Genetic Variation in Hawaiian Drosophila* (1975), 68; *Karyotypic stability and speciation in Hawaiian Drosophila* (1967), 148, 173; *The Species as a Field for Gene Recombination* (1957), 2, 12, 138, 158, 171, 201
Centner, M. R., x, 21, 124, 175–76
Cheyney, J., 42
Clarke, A. E., 146, 150
Clayton, F. E., 148, 173
Cleve, H., 71
Cleveland, L. R., 195
Cody, M. L.: cited by Loftus-Hills, 78
Coluzzi, M., 69
Connell, J. H., 151
Coope, G. R., 77, 81, 87, 151–52, 162, 164, 186
Cooper, M. M., 68, 202
Craig, A.J.F.K., 151–52, 165, 188
Cronquist, A., 51, 172
Crosby, J. L., 4, 14, 23, 25, 28, 84–85, 113–14, 126, 189
Crossley, S. A., 25–26, 113, 125, 128, 131, 134
Crowson, R. A., 172

Daly, J. C., 200, 204
Dana, J. D.: quoted from Ellegard, 94
Darlington, C. D., 53, 59–61, 143, 189
Darwin, C., 39, 60, 64, 76, 93, 95, 97–99, 110, 137, 140, 142, 147, 152, 169–71, 174, 177, 182
David J. R., 66–67
Davidson, G., 204
Dawson, G.W.P., 49–50
Dawson, J. W.: quoted from Ellegard, 94
Dessauer, H. C., 71
Dickerson, R. E., 71

Dobzhansky, T.: *A critique of the species concept in biology* (1935), 1, 32, 138, 140, 142, 201; *An experimental study of . . . Drosophila* (1938); 26; *Genetic natue of species differences* (1937), 2, 5, 6; *Genetics and the Origin of Species* (1937), 2, 99, 137–38, 142–43; ibid. (3d ed., 1951), 6, 22, 32, 41, 65, 99, 113, 138, 142, 144, 159, 181, 202; *Genetics of the Evolutionary Process* (1970), 3–4, 6, 36, 38, 41, 52, 64–65, 80, 88, 119, 139, 141–42, 150, 159; *Humankind: A Product of Evolutionary Transcendence* (1977), 195; *Mendelian populations and their evolution* (1950), xvii, 141; *Organismic and Molecular Aspects of Species Formation* (1976), xvii, 27, 32, 38, 80, 99, 141, 159, 175; *Speciation as a stage in evolutionary divergence* (1940), 14, 15, 26, 41, 80–82, 141–42; *Species of* Drosophila (1972), 40; *Spontaneous origin of an incipient species . . .* (1966), 51
Drew, R.A.I., 208

Ehrman, C., 66
Ehrman, L., 25, 51, 76, 125, 129
Eiseley, L., 95
Eldredge, N., 76, 81, 87, 151, 163, 188
Ellegard, A., 94–95, 143, 169
Erickson, R., 147

Fisher, R. A., 2, 47, 82, 113, 142
Flew, A., 88–89
Fouquette, M. J., 79; cited by Littlejohn, 78
Frost, P.G.H., 9
Futuyma, D. J., 57, 65, 69, 82, 172, 197

Gartside, D. F., 71
Geis, I., 71
Ghiselin, M. T., 88, 137, 146
Gillespie, J., 66
Girard, P., 66
Gordon, D. H., 71
Gould, S. J., 81, 87, 110, 151, 163, 188
Grant, V., 1, 36, 64, 125, 171, 189, 197; cited by Loftus-Hills, 78
Green, C. A., 69, 71
Greenwood, P. H., 148
Gregg, P., 204
Gregg, T. G., 1
Griffiths, G.C.D., 37
Grimstone, A. V., 195
Grine, F. E., 152, 165
Gulick, J. T., 142

Hahn, J., 27
Haldane, J.B.S., xv
Halliday, T. R., 183
Hardy, D. E., 68, 208
Harper, A. A., 21, 114, 126, 176
Harrison, R. G., 69–70, 204
Hartl, D. L., 197
Hedgecock, D., 22–23, 25–27, 78, 83, 202
Hempel, C. G., 118
Henderson, N. R., 21, 163
Hertig, M., 51
Heslop-Harrison, J., 174
Heslop-Harrison, Y., 174
Holstein, M. H., 204
Horak, M., 204
Horsfall, W. R., 186
Hoy, R. R., 27
Hull, D., 98, 153, 177
Hulley, P. E., 151–52, 165, 188
Hunt, R. H., 69

Irving-Bell, R. J., 16, 51, 150

Jackson, J. F., 131
Jackson, R. C., 65
Jaenike, J., 82
John, B., 59, 144
Johnson, W. E., 68, 202
Jordan, D. S., 40
Jordan, K., 40
Jukes, T. H., 191

Kalmus, H., xvi, 67
Kaneshiro, K. Y., 68, 202
Kernaghan, R. P., 51
Kessler, S.: cited by Littlejohn, 84
Key, K.H.L., 23, 42–43, 126, 201
Kimura, M., 191
King, J. L., 191
King, M. C., 71
Kitagawa, O., 26, 66–67
Knight, G. R., 25, 125; cited by Littlejohn, 84
Knox, R. B., 146, 150, 174
Kojima, K. I., 66
Koller, P. C., 26
Koopman, K. F., 25, 113, 125; cited by Littlejohn, 84
Kuhn, T. S., 98, 143, 153

Lack, D., 26, 153
Lambert, D. M., ix, 21, 69, 114, 124, 126, 142, 158, 163, 175–76, 202
Lande, R., 44
Laven, H., 16, 176
Levin, B. R., 196

Levin, D. A., xvii, 8; cited by Loftus-Hills, 78
Lewis, H., 25, 44, 84, 172–74, 176
Lewis, K. R., 59, 144
Lewontin, R. C., 67, 110, 125
Li, C. C., 23, 43, 172, 176
Liebherr, J., 8, 22
Linnaeus, C., 95, 96
Littlejohn, M. J., 2, 26, 76, 78–79, 82–86, 88–89; cited by Loftus-Hills, 78
Loftus-Hills, J. J., 26, 65, 78, 79, 82, 131, 142
Lorentz, K., 184
Lyell, C., 169; quoted from L. G. Wilson, 137
Lyons, N. F., 71

McArdle, B. H., 21, 124
MacArthur, R. H., 22, 125
Macbeth, N., 88
Macnamara, M., 104, 147, 159
Mahon, J. A., 203
Mahon, R. J., 69, 203
Margulis, L., xv, 190, 195–97
Marsh, P. M., 208
Martin, A. A., 82
Maslow, A., 154
Masters, J. C., ix, 142
Mathews, R. W., 21, 27
Maxson, L. R., 70
Mayer, G. C., 57, 65, 69, 82
Maynard Smith, J., 150–51, 163, 191, 196
Mayr, E., *Animal Species and Evolution* (1963), xiii, xvi, xviii, 1–2, 4–5, 20, 22, 26, 33, 36–38, 41–42, 46, 49, 51–52, 64–65, 78, 82, 88, 104, 107, 130–31, 137, 139–40, 142, 144, 148–50, 153, 159, 170–75, 177, 186–87, 189, 201–2; *Change of Genetic Environment and Evolution* (1954), 187; *Difficulties and Importance of the Biological Species Concept* (1957), 49; *Evolution and the Diversity of Life* (1976), 138; *The Evolutionary Synthesis* (1980), 139, 195, 197; *The Growth of Biological Thought* (1982), 171, 197; *Populations, Species and Evolution* (1970), 1, 4, 36, 38, 139, 159, 163, 170, 172; *Principles of Systematic Zoology* (1963), xiii; *Sibling or Cryptic Species among Animals* (1976), 139, 190; *Speciation and Systematics* (1949), xvi, 141, 159, 161, 175; *Speciation phenomena in birds* (1940), 38; *The Species as Category, Taxon and Population* (1987), 187–88; *Systematics and the Origin of Species* (1942), xvi, 22,

40, 137, 139, 142, 170–72, 175, 187, 202
Mecham, J. S., 3, 172
Meeuse, B.J.D., 203
Mettler, L. E., 1, 25, 84
Michod, R. E., 196
Miethke, P. M., 203
Miles, S. J., 69, 150
Miller, R. D., 197
Moore, J. A., xviii, 4–5, 14, 22, 26, 41, 65, 81–83, 88–89
Moran, C., 23, 25, 43, 126
Moran, V. C., 207
Morris, D., 33, 146, 160
Morton, E. S., xvii, 33, 184–85
Muirhead-Thomson, R. C., 203
Muller, H. J., 5, 129
Munroe, E., 69–70, 204
Murray, J., 2, 4
Mutuura, A., 69–70, 204

Nair, P. S., 68
Nevo, E., 67, 71
Noble, G. K., 77
Norris, K. R., 182

Packard, A., 146
Palabost, L., 66
Paley, W., 93, 94
Passmore, N. I., 86, 89
Paterniani, E., 25, 125
Patterson, J. T., 1, 26
Paul, R. C., 27
Pavlovsky, O., 51
Petersen, W., 21
Petit, C., 26, 66–67
Pinese, B., 204
Popper, K., 37, 118
Powell, J. R., 70
Prakash, S., 71
Provine, W. B., 139, 195, 197

Ralin, D. B., 79; cited by Littlejohn, 78
Raven, P. H., 173
Ray, J., 96
Richmond, R. C., 22–23, 25–27, 78, 83, 202
Ringo, J. M., 32
Robertson, A., 25, 125; cited by Littlejohn, 84
Robertson, H. M., ix
Robinson, J. H., 102, 161
Robson, G. C., 142
Roelofs, W. L., 8, 22, 69–70, 204
Rogers, J. S., 71
Rosenberg, A., 191

Sagan, D., 190, 196
Scheemaeker-Louis, D. de, 66
Scoble, M. J., 189
Sene, F. M., 68
Shaw, D. D., 23, 25, 43, 126
Shaw, R. C., 71
Siegfried, W. R., 26
Simpson, G. G., 54
Skolimowski, H., 97, 100
Sleigh, M. A., 196
Smith, D. A., 205
Smith, J. J., 71
Smith, L. S., 205
Sokal, R. R., 202
Spencer, H. G., 21, 124
Spieth, H. T., 68
Stalker, H. D., 148, 173
Stauffer, R. C., 174
Stebbins, G. L., 2, 50, 64
Steiner, W.W.M., 68, 202
Stitch, H. F., 21, 27, 33
Stone, W. S., 1, 26, 68

Takamura, T., 26, 66–67
Takanashi, E., 66
Tardrew, C.: cited by Annecke and
    Moran, 207
Teissier, G., 66–67
Templeton, A. R., 11, 32–33, 189
Tets, G. F. van, 184
Thielke, G., 26, 78
Tinbergen, L., 146
Tinbergen, N., 203
Tobari, Y. N., 66
Tracey, M. L., 22–23, 25–27, 78, 83,
    202
Turner, B. J., 71

Ullyett, G. C.: cited by Annecke and
    Moran, 207

Val, F. C., 68
Van Valen, L., 81, 87, 151, 164
Varossieau, W. W., 203
Vaurie, C., 26
Vawter, A. T., 69–70, 204
Verrell, P. A., 76
Vrba, E. S., 87, 151

Waddington, C. H., 25, 125; cited by
    Littlejohn, 84
Wagner, M., 40
Walker, T. J., 26, 66
Wallace, A. R., 8, 12, 15, 41, 98, 142
Wallace, B., 25, 84, 125
Walter, G. H., 151–52, 165, 188
Walters, C. R., 71
Watson, G. F., 82
Weller, M. W., 153
White, M.J.D., 35–36, 38, 40, 42, 44–
    48, 50–54, 58, 60, 64, 80, 111, 148,
    159, 173, 187
Whitten, M. J., 23
Whittle, C. P., 204
Wiens, J. A., 151
Williams, G. C., xvi, 3, 6, 53, 64, 79–80,
    88, 98, 101, 109–10, 139, 146, 160,
    170, 172, 182
Wilson, A. C., 70–71
Wilson, D. S., 107
Wilson, E. O., 22–23, 25, 27, 83, 113,
    125–26
Wilson, L. G., 137
Wollaston, T. V.: quoted from Ellegard,
    94

Yen, J. H., 16
Young, J. Z., 143

Zaslavsky, V. A., 126

Subject Index

Adaptive device: isolating mechanism as, criticized, 28, 38, 100, 159
Adaptive entity: species as, criticized, 3
"Adaptive polymorphism," 107
Allopatric speciation, 8–9, 22, 40, 165, 176; "Class I" (empirical) evidence for, 3, 37, 115, 121, 140; difficulty for isolation concept of species, 119, 142, 159; pleiotropy and, 141–42

Biological species concept, 38, 105, 137, 158. *See also* "Evolutionary synthesis"; Isolation concept of species
Biparental organisms: discussion restricted to, 2, 138; fate of polyploid among, 173; fertilization between, in normal habitat, 147, 160, 182; mate recognition in all, 202; need mechanism for fertilization, 100–101, 105, 144; recognition concept applies to all, 120, 144; sessile and motile, 120, 146. *See also* Sex
Blyth, E.: influence on Darwin, 95; sterility for species preservation, 99

"Catastrophic selection" (saltational speciation), 44
Chromosome: introgression in parapatry, 43; "races," 43; rearrangement, correlate not cause of speciation, 44–45, 69; segregation distorted (meiotic drive), 42, 47, 115; "semipermeable filter" of, 43
Competition: ecosystem and, theory, 116, 152; "ghost of competition past," 15, 41; interspecific, as evolutionary motor, 151–52; intrasexual, for mates, 76, 183; nonreliance on, in recognition concept, xix, 151, 165
Computer modeling: maintenance of parity in, 85; and predictions of deterministic model, 25, 113, 176; to simulate reinforcement, 4, 25, 176; utility of, 112–14
Conflation: defined following Wittgenstein, 111; generates nonsense, 111; between the recognition and isolation concepts of species, xvi, 171, 195; of species

concepts in taxonomy and population biology, 36, 140, 158, 189, 200
Creation stories, 93, 95–97, 143; in Genesis, 93, 95, 142; of man, 96, 98–99, 137; Paley's Grand Watchmaker, 93; Ray's belief in great design for continuance of species, 95; Xenophanes of Colophon, 102. *See also* Species concepts, speciation
Cryptic species, 69, 101, 117, 163, 199–209; case histories of, 203–5; and control campaigns, 207; correlation with cytology, 206; correlation with proteins, 205–6; detection of, 68, 199–209; electrophoretic studies of, 205; the isolation concept and, 202; of parasitic Hymenoptera, 206–7; the recognition concept and, 202; in taxonomy, 202
Cytoplasmic incompatibility, 16, 51, 150; in the *Culex pipiens* complex, 16; an intraspecific phenomenon, 18, 176. *See also* Sterility

Darlington, C. D.: on G. E. Allen's book, 60; on C. Darwin, 60; on T. H. Morgan, 60; obituary, 58–62; on M.J.D. White's book, 60
Darwin, C.: on species and speciation, 92–102. *See also* Sterility
Delimiting genetical species, 2, 105; a consequence of the fertilization system, 147–48; by preventing gene exchange, 139–40, 153
Dendrogram of relatedness: based on set amount of genetic change, 64. *See also* Genetic distance
Directional selection, 101

Ecology, species concepts in, 158–65
Effect, definition of, 109–10
Electromorphic distance and speciation, 63
Environment. *See* Habitat; Recognition concept of species; Red Queen hypothesis
"Escape from hybridity," 53, 189
Evidence: for allopatric speciation, 37, 115, 121, 140; "Class I" type, defined,

Evidence—(*continued*)
37; "Class II" type, defined, 37; for
isolating mechanisms, 4; for speciation
by reinforcement, 8
Evolution: of eukaryotic cell, 195;
idealistic views of, 96–97; of mitosis,
195; neutral theory of molecular, 191;
nonteleological view of, 97; of sex,
195–96
"Evolutionary synthesis," 180–81, 190–
91, 194–98; and the biological species
concept, 190; as opposed to the term
"neo-Darwinism," 194; updating, 194–
98
Evolutionary theory: conflation between
different, 111; dangers of verbal
models in, 113; intellectual honesty in,
93; modeling in, 77, 128;
preconceptions behind, 100; rigor of,
88–89; and semantics, 106; teleology
in, 108–9

Fertilization: between motile and
between sessile organisms, 146;
significance of, 33; in specific habitat,
13. See also Fertilization system
Fertilization mechanisms. *See*
Fertilization system; Recognition
concept of species; Specific-mate
recognition system
Fertilization system, 105, 202; characters
of, and natural selection, 116, 146–48,
184 (*see also* Fertilization mechanisms;
Natural selection; Specific-mate
recognition system); common, among
members of a species, 105, 147, 150,
196; definition of, 175, 183, 201–2;
function of, 144, 182–83; influence of
way of life on, 120, 144, 146; of man,
116, 182; the same, in different
populations, 184, 208; and speciation,
185; stability of, 161, 163, 184–85;
and test of the recognition concept,
118, 52, 161–62. See also Recognition
concept of species, fertilization
mechanisms
"Field for gene recombination," 159;
essential property of a species, 2, 138,
158–59, 201; evolution of a new, 148
(*see also* Speciation); isolation concept
and pleiotropic change of limits to, in
allopatry, 141, 160; limits to, 12, 105,
138–40, 147; and man, 153–54. *See
also* Gene pool; Species concepts
"Flush-crash" cycle, 70
Function: definition of, 109–10; of
fertilization system, 144, 182–83; of

isolating mechanism, 8; Williams'
definition of, 6. *See also* Effect

Gene pool: defined, 2; integrity of, an
isolationist view, 2–3, 37, 189;
polyploidy, 50, 149; 173; protection
of, by isolating mechanisms, 3–4, 37,
43, 105; selection acting on rare
elements in, 23, 25. *See also* "Field for
gene recombination"
Genetical species concept, 2, 52, 137,
201; delimiting factors in, 2, 105,
139–40, 147–48, 153; isolationist
understanding of, 171; recognitionist
understanding of, 171; versus
taxonomic concept of species, 63–72.
See also "Field for gene recombination"
"Genetic coupling," 27
Genetic distance, 64–65; dendrogram of
relatedness based on set amount of,
64; gene rearrangement and, 68;
measures of, 63–72; morphological
difference and, 68; set amount of, 64;
and speciation, 63; taxonomy and, 68;
and time after speciation, 72
"Genetic revolution," 47; prerequisite
for change of habitat, 186. *See also*
Speciation theory, criticisms of
sympatric
Geographic speciation. *See* Allopatric
speciation
Gradualism, 121, 165, 188. *See also*
Punctuated equilibria
Group selection: and "adaptive
polymorphism," 107; a consequence
of the isolation concept, 107, 152; in
contrast to individual selection, 107,
159; invoked if species are adaptive
devices, 80

Habitat: hybridization of, 184
—normal (preferred, recognized):
adaptation to, 53, 55, 86; central to
the recognition concept, 164; change
of, and speciation, 33, 43, 161–62,
164; efficiency of SMRS signals in, 33,
55, 80, 82, 85–86, 89–90;
environment in which a species
evolves, 158–65, 182; fertilization in,
65, 80, 86, 147, 160–61, 184;
"recognized habitat," 101–5; stability
in, 81, 87, 123, 152, 164; tracking of,
72, 83, 87, 91, 151, 162, 186 (*see also*
Red Queen hypothesis)
Heterozygous disadvantage: in *Clarkia*,

176; elimination of cause of, 84; unstable polymorphism and, 14
"Host races" and sympatric speciation, 47–48
Hybridization: artificial conditions and, 184; change in frequency of, 26, 113; consequence of, varied, 162; of the habitat, 184; heterozygous disadvantage after, 175; "integrity of species" corrupted by, 95–96, 99, 101; interspecific, 5; "macrorecombination" another term for, 194; natural selection not involved in preventing, 5; preconceptions when theorizing about, 96–97, 189; race between character displacement and, 126; reproductive isolation and, 8; and speciation, 197
Hybrid sterility. *See* Sterility
Hybrid zones, 126; artificial and experimental, 44, 81; gene migration away from, xvii–xviii, 129 (*see also* Moore's objection; Parapatric zone); of morabine grasshoppers, 43; parapatric, 43; selection for positive assortative mating in (*see* Positive assortative mating; Speciation by reinforcement); as a "semipermeable filter," 43, 126

Individual selection, 107, 159
Isolating mechanism. *See* Isolation concept of species
Isolation concept of species. *See also* Recognition concept of species; Species concepts; Taxonomic concept of species
—consequences: contradictory postmating isolating mechanism, xvii, 3, 8, 18, 39, 172; difficulty with allopatric speciation, 118–19, 142, 159; no falsification of negative heterosis prediction, 25, 142; flaws, 118; group selection, 80, 107, 152 (*see also* Group selection); influence on evolutionary thought, 1–9; logical, 39; positive recognition versus negative isolation, xvi, 5, 9, 21, 65; stabilizing selection away from hybrid zone, 15 (*see also* Moore's objection)
—idea: ambivalence, 5; culture, 99; Dobzhansky, 138–40, 142; genetic integrity and, 159, 201; historical roots of, 143; inconsistency in, 159, 170; Mayr, 138–40; as paradigm, 98–100, 143, 153; and plants, 173–74; relational aspect of, 105, 139; versus

taxonomic concept of species, 36–37
—isolating mechanism, 99, 139, 201; as adaptive device, 159; as ad hoc mechanism, 41; criticisms of, 3; as effect, 3, 9, 79, 101; evidence for, 4; function of, 8; heterogeneity of, 5, 172; loss of, in allopatry, xviii; negative aspect of, 2; postmating, 3, 172; premating, xvii, 172; problem of allopatric origin, 142; properties, 39; and sexual dimorphism, 153; and sterility, 175; the term, 1–9; types of, 39
—speciation: chromosomal models of, 42; critique of sympatric, 45; inferring causality from correlation, 173; "instant," 49, 115, 149, 170, 172–73; and pleiotropy, xvii; by polyploidy, 149, 172; by reinforcement (*see* Speciation by reinforcement); "species integrity," 28; stasipatric, 42–44
—species: as adaptive entities, 3; the biological species concept, 38, 105, 137, 158

Macrorecombination, 194–98, *See also* Hybridization
Mate: recognition of (after pairing), 76; "selecting mechanism," evolution of, in noisy environment, 86. *See also* Isolation concept of species
"Mate choice," 86; criticism of term, 76, 87; not implied in recognition concept, 79, 101, 127, 146, 202
"Mate recognition," criticism of term, 76
"Mate selection," criticism of term, 79
"Meiotic drive," 42, 47, 115
Modeling: logical rules for, 37. *See also* Computer modeling
Moore's objection, xvii–xviii, 14, 81, 88, 129. *See also* Hybrid zones
Morphological resemblance and genetic distance, 63–72
Morphological variation and species stability, 67

Naive panselectionism, 148
Natural selection: action of, if negative heterosis, 23–25, 84–85, 125, 150, 175; action of, in changing environment, 87, 151 (*see also* Red Queen hypothesis); against hybrids, 5, 21, 189; direct role of, in speciation (speciation by reinforcement), xvii, 3–4, 12, 22, 39, 41, 141–42; in evolution of isolating mechanisms, 3, 6–8, 28, 110–13; in evolution of recognition

Natural selection—(*continued*)
mechanisms, 5, 33, 65–67, 101, 116,
146–48, 184 (*see also* Fertilization
mechanisms; Specific-mate recognition
system); for hybrid viability, 189–90;
outcome of, in computer simulations,
131; role of, in maintaining "specific
integrity," 2, 95, 97, 189; and self-
incompatibility, 174; and sterility, 99,
170. *See also* Selection, stabilizing
Negative heterosis: algebraic
considerations of, in population
genetics, 23, 43, 124–26; in *Clarkia*,
176; computer simulations of, 25;
elimination of cause of, 84;
experimental simulation of, 25, 124–
26; natural examples of, 25, 43–44;
and speciation by reinforcement, 25.
See also Heterozygous disadvantage
Neo-Darwinism. *See* Evolutionary
synthesis
Niche: characteristics stable throughout
species' range, 152; speciation and
acquisition of a new, 148; theory
and species, 39–40; 116, 152;
"unoccupied," 188–89

Parapatric zone: action of natural
selection at, 126; chromosome
introgression across, 43, 129, 206;
gene migration away from, xvii–xviii,
129 (*see also* Moore's objection);
hybridization at, 43; reproductive
character displacement at, 26, 140,
206 (*see also* Speciation by
reinforcement); speciation at, 39; of
Vandiemenella viatica, 43; versus
sympatry, 128. *See also* Hybrid zone
Phyletic gradualism, 121, 165, 188. *See
also* Punctuated equilibria
Plants: differences in ploidy among, 49;
hybrid disadvantage and, 16;
hybridization between, when habitat
changed, 184; hybrid origin of
tetraploid, 50, 189; importance of
Grant's *Plant Speciation*, 36;
intersterility and population studies
(*Clarkia*), 25; isolation concept and,
120, 174; rarity of autopolyploid, in
nature, 50; recognition concept and,
120, 138, 146–47, 168; self-
incompatibility, 61, 150, 174;
simulated reinforcement situation, 4,
85; sterility in, 51, 173, 176;
taxonomic problems arising from
polyploidy in, 50–51, 172
Polymorphism, "adaptive," 107; a group

character, 107–8; unstable, if
heterozygous disadvantage, 14
Polyploidy, among biparental organisms,
173; and the gene pool, 50, 149, 173;
taxonomic treatment of, 50–51, 172
Popper, K., on "ad hoc shoring up," 118
Population genetics. *See* Isolation
concept of species, speciation;
Recognition concept of species,
speciation
Positive assortative mating: alleles for,
15–16; artificial increase of, 25–26,
113, 125, 128, 131, 134; change in
degree of, measured, 124–31; change
to, affects mate recognition, 16; effect,
not function, of SMRS, 9, 37, 79,
100–101, 146; evidence for increased,
26, 125; dynamics of changes to, in
simulation studies, 124–31; selection
for, 4, 14–15; testing for, among
allopatric populations, 208
Postmating isolating mechanism, 3, 172;
inherent contradiction of the term,
xvii, 3, 8, 18, 39, 172. *See also*
Isolation concept of species
Preferred habitat. *See* Habitat, normal
Premating isolating mechanism, xvii,
172. *See also* Isolation concept of
species
Punctuated equilibria, 81, 87, 116, 121,
151, 163, 188. *See also* Phyletic
gradualism

Recognition: definition of, 101, 146;
mate, 5, 76; of mate after pairing, 76;
species (*see* Species recognition, a
human activity); versus choice, 79,
101, 127, 146, 202
Recognition concept of species. *See also*
Isolation concept of species; Species
concepts; Taxonomic concept of
species
—definition of, 147, 161, 201; early
definition of, 21; essential aspect of,
37, 80; a test of, 118
—consequences, 104–21, 136–54, 161–
63, 187; advantages in biology, 120–
21; for applied biology, 199–209; for
ecology, 116, 188–89; importance of
sex in, 143–44, 160; individual
selection, 107; nonrelational, 105,
149; philosophical, 104–21; for
population biology, 116; positive
assortative mating, 79 (*see also* Positive
assortative mating); positive
recognition versus negative isolation,
xvi, 5, 9, 21, 65; of sterility, 175;

paleontological, 151, 188; stability, 65–66, 87, 152
—fertilization mechanisms (SMRS), 32, 80, 100, 160, 184, 203; behavioral 6, 12; chemical, 69–70; in *Clarkia* 176; coadapted signal and response, 163; correct perspective of, 32–34; in diurnal and nocturnal insects, 203; evolution in normal habitat, 86; of *Homo sapiens*, 154; of monotypic groups, 153–54; of plants, 120, 146–47, 174 (*see also* Plants); and sexual selection, 76; stability, 9, 21, 163, 185; types of, 100; ultrasonic, 71; utility of optical, in taxonomy, 68; variation of, 27; and way of life, 160
—habitat and way of life, 80, 85–86, 158–65, 175–76; ecosystem structure, 152
—speciation, 27, 33, 66, 147–49, 185–86; in allopatry, 176; chromosomal rearrangement, 44; "Class I" evidence, 37, 114, 140; "Class II" evidence, 37, 114 (*see also* Evidence); correlation with chromosome fixation, 148–49; critique of sympatric, 45–46; ecological implications, xviii; empirical evidence for allopatric, 115; environment, 158–65, 186; importance of chance, 116; as incidental consequence of adaptation, xvii, 176; and small populations, 101; without competition, xix (*see also* Competition)
Red Queen hypothesis, 81, 87, 91, 116, 151, 64
Reinforcement, within a species, 78. *See also* Isolation concept of species; Speciation by reinforcement
Reproductive character displacement, 78; at parapatric zone, 26, 140, 206; race between hybridization and, 126. *See also* Speciation by reinforcement
Reproductive community, 53
Reproductive environment, 85
Reproductive isolation. *See* Isolation concept of species

Selection: "catastrophic" (saltational speciation), 44; directional, 101; group (*see* Group selection); individual, 107, 159; natural (*see* Natural selection); sexual (*see* Sexual selection); stabilizing, 15
Self-incompatibility in plants, 61, 150, 174
Semantics, and species concept, 106,

176. *See also* Evolutionary theory
"Semipermeable filter" (of chromosomes), 43
Sex: alternatives to, xv; in eukaryotes, 144; evolution of, xv, 181, 190, 195; evolutionary advantages of, 196; and the origin of first species, 197; physiological synchronization and, 13; and species concept, 2, 181
Sexual characters, secondary, 182
Sexual selection, 76–77, 79, 183; and SMRS, 76; unstable, due to positive feedback, 82
Sexual signal. *See* Signal, sexual
Sibling species. *See* Cryptic species
Signal: chemical, 6, 100; coadaptation, 27; environmentally adapted, 101; interference, 85–86; in noisy environment, 86; pheromone, 8; sexual, 6, 13, 86; shared between species, 184
SMRS (specific-mate recognition system). *See* Recognition concept of species; Specific-mate recognition system
Speciation, allopatric. *See* Allopatric speciation
Speciation by reinforcement: an adaptive process, xvii, 27–28; bidirectional selection, 14; closed population model, 83–84; criticisms of 1, 20–28, 76–89; experimental flaw, 84; experimental studies and simulations, 3–4, 25, 84–85, 124–31, 176; "ghost of competition past," 15, 41; instability after artificial selection for, 85; lack of evidence for, 8, 131; Moore's objection, xviii, 14, 81; necessary conditions, 113–14, 124–31, 173; parapatric interface, 26; purported examples of, 26; secondary overlap, 125; sexual dimorphism as evidence for, 26; sterility, 171; time required for, 14; typical statement of, 22. *See also* Isolation concept of species, speciation; Negative heterosis
Speciation theory: ad hoc shoring up of, 114; criticisms of sympatric, 45–48, 150; "host races" and sympatric, 47–48
Species. *See* Isolation concept of species; Recognition concept of species; Species concepts; Taxonomic concept of species
Species, sibling. *See* Cryptic species
Species, taxonomic. *See* Taxonomic concept of species

"Species," the term, as a homonym, 163
Species concepts. *See also* Isolation
concept of species; Recognition
concept of species; Taxonomic
concept of species
—bias: ad hoc shoring up of, 118, 198;
and auxiliary hypotheses, 118–9;
Christian view, 169; conflation, xvi,
36, 111, 195, 200; semantics, 176;
teleology in, 152–53
—evolutionary importance: in economic
entomology, 199–209 (*see also* Cryptic
species); implications arising from,
xviii, 2; in paleontology, 158, 162; in
population genetics, 36; in taxonomy,
36, 199–209
—history of: and culture, 142; biological
species concept, 38, 105, 137, 158 (*see
also* Isolation concept of species);
Lyell, 116, 137; and traditional views,
169
—kinds: distinguished, 104; genetical, 2,
52, 137, 171, 201; sexual and asexual,
120; taxonomic, 2, 52. *See also* "Field
for gene recombination"
—speciation: allopatric, 8, 22, 40, 115,
159, 165, 176; argument for, in large
populations, 115; argument for, in
small populations, 114; change of
habitat and, 164; creationist views of,
93–97; Darwin's views of, 92–102;
different modes of, 187; difficulty
detecting sympatric, 48; genetic
change during, 67, 70; hybrid origin,
50; importance of chance, 116;
influence of population size on, 185;
pattern of, in paleontological record,
165; protein differences after, 71;
saltational, 44; theories of, 22
Species diversity: causes of, xv; natural
selection and, 12; radiation and, xv;
recombination and, xv; role of
selection in, 3
Species habitat. *See* Habitat, normal
Species hybridization, prevention of, 5.
See also Hybridization
"Species integrity," an isolationist view,
2–3, 37, 80, 189
"Species recognition," 5; criticism of

term, 76; a human activity, 76
Specific-mate recognition system
(SRMS), 1, 8; not "mate recognition,"
144; not "species recognition," 144,
146; origin of the term, 1; in plants,
147 (*see also* Plants); selection and, 86;
as subclass of fertilization mechanism,
80. *See also* Recognition concept of
species, fertilization mechanisms
Stabilizing selection. *See* Selection,
stabilizing
Sterility, 51, 150; as a "barrier," 18;
cytoplasmic incompatibility and, 16,
51, 150; Darwin on hybrid, 170; as an
effect, 97, 99, 110; problem of,
between species, 168–77; species
defined in terms of, 51, 168–77; for
species preservation, 99. *See also*
Cytoplasmic incompatibility; Isolation
concept of species, isolating
mechanism
Subspecies, 42
Symbiosis, 194–98
Sympatric speciation. *See* Speciation by
Reinforcement
Syngamy and chance, 13

Taxonomic concept of species, 163, 200;
asexual organisms, 52, 189; asexual
speciation and, 52; clones, 53; and
diploid number, 71; and G-banding
pattern, 71; parthenogenesis, 201;
polyploidy and, 172; versus genetical
species concept, 63–72
Taxonomy: view of polyploidy in, 172

Wallace, A. R.: influence on current
reinforcement theory, 12; theistic
inclination, 98
Way of life: influence on fertilization
system, 80, 85–86, 120, 144, 146,
158–65, 175–76, 184
Williams, G. C.: "on adaptation," 3, 6;
Adaptation and Natural Selection, 109–
11; definition of "effect," 8, 109–10,
160; on "function," 6, 109–10; on
"mechanism," 6; on nature of species,
98, 110